RAND NATIONAL DEFENSE RESEARCH INSTITUTE

T0289473

Improving DLA Supply Chain Agility

Lead Times, Order Quantities, and Information Flow

Eric Peltz, Amy G. Cox, Edward W. Chan, George E. Hart,
Daniel Sommerhauser, Caitlin Hawkins, Kathryn Connor

Prepared for the Office of the Secretary of Defense

For more information on this publication, visit www.rand.org/t/RR822

Library of Congress Cataloging-in-Publication Data is available for this publication.
ISBN: 978-0-8330-8866-6

Published by the RAND Corporation, Santa Monica, Calif.
© Copyright 2015 RAND Corporation
RAND® is a registered trademark

Cover photo: U.S. Air Force/Master Sgt. Jon Nicolussi (Fotolia © learchitecto)

Support RAND
Make a tax-deductible charitable contribution at
www.rand.org/giving/contribute

www.rand.org

Preface

The Defense Logistics Agency (DLA) supplies common military items to the armed services and other organizations. In doing so, it faces the challenge of maintaining a high degree of customer service at minimum cost. One cost that DLA faces is buying inventory that becomes excess and is disposed of because demand for the materiel has unexpectedly declined. On average, DLA disposes $1 billion of inventory every year, which represents $1 billion in spending that was ultimately not needed. When customers' needs for stocked items change, DLA responds by purchasing more or less of the items, as appropriate. However, when demand for an item unexpectedly declines with less warning time than the order lead-time horizon, excess inventory is likely to develop. Similarly, when demand for an item unexpectedly increases—again, with less warning time than the order lead-time horizon—stockouts occur while waiting for new orders to arrive. Improved *supply chain agility*—the ability to respond quickly and efficiently to changes in demand and supply—would mitigate the effects of these demand shifts, reducing costs and improving customer service. This report examines how DLA could improve its supply chain agility, focusing on response to demand changes from the high dynamism in demand that DLA faces. Naturally, this involves many aspects of improving supply capabilities in addition to improved demand change awareness and sensing. Specifically, the report focuses on reducing lead times, reducing and right-sizing order quantities, and mitigating the effects of demand shifts through improved customer information flow and utilization.

This research was sponsored by the Assistant Secretary of Defense for Logistics and Materiel Readiness, in coordination with DLA Logistics Operations (J3) and Acquisition (J7). It builds on earlier RAND research that found that lead times and order quantities are the most important drivers of DLA inventory management performance because of the stockouts and excess inventory that result from changes in demand when combined with relatively long lead times and large order quantities.[1] This report should be of interest to people working in DoD supply chain management, along with other stakeholders in DLA's supply chain management effectiveness.

---

[1] E. Peltz and M. Robbins, with G. McGovern, *Integrating the Department of Defense Supply Chain*, Santa Monica, Calif.: RAND Corporation, TR-1274-OSD, 2012.

This research was conducted within the Acquisition and Technology Policy Center of the RAND National Defense Research Institute (NDRI), a federally funded research and development center sponsored by the Office of the Secretary of Defense, the Joint Staff, the Unified Combatant Commands, the Navy, the Marine Corps, the defense agencies, and the defense Intelligence Community. For more information on the RAND Acquisition and Technology Policy Center, see http://www.rand.org/nsrd/ndri/centers/atp.html or contact the director (contact information is provided on the web page).

Contents

Figures

Tables

Summary

The core mission of the Defense Logistics Agency (DLA) is to efficiently support the armed services' requirements. To fulfill this mission, it must maintain inventories of items for which demand is highly variable even when relatively stable, could increase or decrease dramatically on short notice, or may never even materialize. While commercial firms might consider stocking such items unprofitable and would be unlikely to do so without a purchase guarantee, DLA does not have this option if it is to effectively support the warfighter. Thus, the high level of customer support that DLA has achieved has come at high inventory cost, manifested in inventory excess that leads to disposals. From 2005 to 2013, DLA disposed of an average of more than $1 billion per year. This represents prior spending for items that were ultimately not needed. In the Department of Defense (DoD), this cost of doing business is charged by DLA to its customers through its cost recovery rate or surcharge that it is mandated to add to material costs when charging its customers for goods in order to cover its operating costs.

To buy inventory and thus support customers effectively, DLA has to anticipate or forecast demand. The longer it takes to procure an item from a supplier—the *lead time*—the longer the horizon over which DLA has to project the likely demand. Similarly, the larger the order size relative to demand, the longer the horizon over which DLA has to project demand. As lead time lengthens, forecasting becomes more and more prone to error because it works well when demand can be projected from trends but breaks down when unanticipated events occur that change the trend. The longer the lead time, the higher the probability is of such an event within the lead-time window for ordering an item. Sudden spikes or large upward shifts in demand can lead to lengthy stockout periods—as could demands for newly needed items. Conversely, downward demand shifts—including ceasing entirely after a change in operational use or equipment—can lead to inventory excess. The frequency and severity of these conditions increase as lead time does.

To reduce the risk of excess inventory and thus reduce disposals and material costs, and to improve customer support by reducing the length of stockouts from surprise demand increases, DLA needs to improve its supply chain agility. *Supply chain agility* is better aligning supply with demand to serve customers as well as or better with less inventory on hand or on order. More generally, we define improving supply chain agility as improving the responsiveness and efficiency with which customers with changing needs are served through strong supply chain alertness and rapid supply chain response to changes in demand and supply, to include changes specific to individual items and those from broader changes in the environment that affect item segments or all or much of the item population.[1] DLA clearly faces very dynamic demand at the item level even in the absence of major environmental changes. This presents a significant challenge for efficient and effective inventory management. The dynamism of supply appears to be less than demand, with major problems from supply-side volatility not apparent or documented in prior research. Given this and prior research that identified a critical need to improve DLA's responsiveness to changes in demand,[2] we focus on this aspect of supply chain agility in this report.

Improvements in three processes enable supply chain agility with respect to handling demand changes: increasing the speed and accuracy of the delivery of information about planned changes in customer demands, shortening lead times, and reducing order quantities.

Based on analyses of DoD and DLA policy, commercial and academic literature, interviews, and inventory and contract data, this report addresses how DLA can improve these three processes. DLA is already implementing some of the recommendations, and progress should be realized soon.

Findings and Recommendations

The central finding and recommendation of this project is a need for increased enterprisewide emphasis on supply chain agility, in essence making it part of DLA's DNA so that it imbues all processes and decisionmaking. This does not mean it should always be the prime factor, but rather that it always needs to be considered. Overall metrics and agility-related metrics, whether for outcomes or processes, must be applied enterprisewide. This is because the processes that interact to determine a supply chain's main outcomes (i.e., materiel availability, inventory turns, and purchase price) are affected by multiple functions and organizations across the supply chain. The only place where all DLA outcomes come together is the DLA director level. However, the services also affect forecasting and lead times, which means that the only place where all of these outcomes and factors come together is at the DoD supply chain enterprise level. Thus, an increased emphasis on supply chain agility requires involvement from the most-

---

[1]   For a useful review of supply chain agility definitions, see D. M. Gligor and M. C. Holcomb, "Understanding the Role of Logistics Capabilities in Achieving Supply Chain Agility: A Systematic Literature Review," *Supply Chain Management: An International Journal*, Vol. 17, No. 4, 2012b.

[2]   E. Peltz and M. Robbins, with G. McGovern, *Integrating the Department of Defense Supply Chain*, Santa Monica, Calif.: RAND Corporation, TR-1274-OSD, 2012.

senior management levels across the DoD supply chain management enterprise and flowing downward to all levels.

Our first recommendation, therefore, is for strong management involvement in pursuing agility across the enterprise at all levels. Such involvement has three elements: ensuring the integration of supply chain agility into all processes through management emphasis, adding new metrics and enhancing some existing ones, and training all stakeholders. Process integration involves not only policy and procedures but also day-to-day engagement—for example, asking agility-related questions in decision briefings and reviews to emphasize agility and continually force all to consider how they can affect it. New metrics include

- using inventory turns for all items in the DLA monthly Agency Performance Review (APR)
- using award price change compared with the producer price index in the APR
- adding administrative lead time (ALT), production lead time (PLT), and order quantity overrides to the Inventory Management Council monthly report
- measuring the services, in conjunction with DLA, on demand forecast accuracy for DLA-managed items.

Additionally, monitoring disposals over time as a percentage of sales and should-be inventory would provide an overall view of whether agility is improving (in conjunction with continuing to measure materiel availability and ensuring that it stays the same or improves). Fully training all communities on supply chain agility would ensure that they understand its importance and how they affect it, enabling them to combine the motivation to improve agility with the knowledge to do so.

Our second recommendation is to continue to shorten lead times. DLA's efforts to reduce ALT have expanded since the start of this project, and we recommend continuing these efforts and implementing the DLA Time to Award team's recommendations. The next step is to expand such efforts to PLT reduction, with this also becoming a charter of the DLA Time to Award team.

For PLT, we recommend that DLA work with its suppliers to track PLT and use it as a continuous improvement metric, with performance improvement favorably affecting supplier selection. DLA should also work with its major suppliers to identify what sizing and pacing of orders would enable the lowest costs and shortest lead times. In addition, we recommend that DLA combine contracts where possible to lower transaction costs for suppliers. Further, when different parts of DLA place separate orders with the same supplier, if one order becomes a critical high-priority need and a stockout has developed or is projected, this need should be discussed and coordinated among the DLA supply chains before conveying that information to the supplier.

We also recommend that DLA use PLT to a greater degree and in a more standard way in bid selection. We found that buyers have some understanding of the importance of PLT but lack the guidance, data, tools, and thorough understanding of supply chain agility to fully address it in their processes. Regarding guidance and policy, this means explicitly defining *best value*, to include lead times and order quantities in addition to price. Regarding data, we recommend that DLA work with Defense Procurement and Acquisition Policy to expand the data that are available in the past performance information retrieval system, the data system on supplier performance, to include PLT. For tools, buyers and supply planners need a tool that weighs the value of competing options by accounting for trade-offs among prices, order quantities, and lead times, particularly when faced with a price break for a bulk purchase. We offer an example of such a tool in the report. Finally, we recommend maximizing the use of items that also have commercial demand, or *dual-use items*, which tend to have shorter lead times, where possible.

Our third recommendation is to *right-size* order quantities for the optimal balance of purchasing workload and inventory levels to minimize total cost. Small order quantities relative to demand minimize the risk of purchases that lead to inventory disposal, especially for items with long lead times. At the same time, processing more small orders requires more acquisition labor. In the legacy DLA system for determining order quantities, the implied inventory holding cost has been putting too much weight on minimizing inventory, in turn leading the DLA supply chains to apply manual overrides to reduce the high workload burden (note that *supply chains* are organizational entities in DLA encompassing all of the functions related to providing material for a product segment).[3] Instead, we recommend using the economic order quantity with an inventory holding cost that closely approximates the estimated DLA holding cost. This would decrease the size of order quantities for items with high annual demand values (ADVs), and it would also decrease the number of purchase requests overall, primarily through reductions among items with low ADVs. Thus, both inventory levels and the number of purchase requests that acquisition staff have to fulfill would become rebalanced. In the longer term, DLA should compute safety stock levels jointly with order quantities in order to optimize inventory fully.

Our fourth recommendation is for DLA to continue to expand its use of long-term contracts (LTCs), especially with guaranteed minimums and longer lengths, and where low-demand and high-demand items can be combined. LTCs are associated with shorter lead times and facilitate smaller order quantities, and most suppliers appear to want more LTCs, especially when accompanied by guaranteed minimum orders and longer contract lengths. Prioritizing National Item Identification Numbers (NIINs) with the highest annual demand value for placement on long-term contracts realizes the greatest benefits.

---

[3] The DLA supply chains are organizational entities encompassing all of the functions related to providing material for a product segment. DLA's supply chains include Aviation, Land and Maritime, Troop Support, and Energy, and the scope of this report is limited to the first three.

Our fifth recommendation is for DLA to work with the services to improve the flow of information from them to DLA about upcoming changes to NIINs. Faster, more-accurate transmission of information about upcoming changes in demand improves supply chain agility. Currently, DLA and the services have processes in place to collaborate on demand planning, especially with the depots. Tighter collaboration would also be valuable when one of the services' engineering organizations institutes an engineering change that results in replacing an item with a new version, requiring the old item to be phased out, and establishing inventory of the new item. DLA can avoid both oversupply of the old item and undersupply of the new item when this information is transmitted effectively and as early as possible from the services to DLA. However, conveying information about engineering changes involves many personnel and steps in a complicated process. As a result, the information can become slowed and the needed recipients confused. Management-level emphasis and process redesign within and across the organizations and training about how individual roles affect supply chain agility can improve this outcome. We recommend developing an integrated process team to examine and improve the processes involved with engineering changes and related information-sharing. Because dual-use items tend to have shorter lead times, the process team's charter could also include examining how service process changes and increased collaboration with DLA could increase the frequency of pursuing dual-use items during engineering change and item development and selection.

Finally, one way to give DLA information at different stages of engineering changes may involve developing a repository for exchanging such information. It could designate the likelihood of change for a NIIN from no change being considered to potential change in development to change likely or in process. DLA could then use this information to improve decisionmaking and reduce risk. For example, identifying NIINs being considered for replacement would enable DLA to scrutinize bulk buys more carefully to avoid purchases of potentially excess inventory. As collaboration between DLA and the services improves, we recommend sharing the resulting improved information with suppliers. To account for variation in suppliers' needs and capabilities, the information can be provided with portals through which suppliers could access forecasts, inventory levels, and other information at their own pace.

These recommendations, which span the supply chain from top to bottom and from end to end, can improve DLA's supply chain agility. The recommendations in this report, though, will not fully exploit the potential of enhanced supply chain agility but rather will represent a strong start to improving it, emphasizing the most urgent areas for DLA to begin or to continue addressing. They reflect a first step on the journey toward enhanced supply chain agility, which promises substantial financial and customer service gains.

Acknowledgments

This research would not have been possible without the assistance and support of many people. First and foremost, we want to thank members of the Defense Logistics Agency (DLA) at its headquarters and at its Aviation, Land and Maritime, and Troop Support supply chains, and members of the military services. We also express our appreciation to the commercial companies, both DLA suppliers and other firms, that shared their knowledge and time. We are grateful to our sponsors, initially the Honorable Alan Estevez, then–Assistant Secretary of Defense for Logistics and Materiel Readiness, and Mr. Paul Peters as Acting Assistant Secretary of Defense for Logistics and Materiel Readiness, for supporting this research. In addition, Mr. Peters provided guidance and support throughout the project. We thank the reviewers of this report, David Gligor and Elvira Loredo, who lent insightful comments that improved the final product. Finally, we thank our RAND colleagues for their dedicated work in bringing this report to fruition, especially Cynthia Cook, Marc Robbins, Cheryl Kravchuk, Judy Mele, Shayla Ball, Danny Tremblay, and Laura McMillen.

Abbreviations

ABVS	automated best value system
ADV	annual demand value
ALT	administrative lead time
APR	Agency Performance Review
CovDur	coverage duration
DLA	Defense Logistics Agency
DLAD	Defense Logistics Acquisition Directive
DoD	Department of Defense
DORRA	DLA Office of Operations Research and Resource Analysis
DPAP	Defense Procurement and Acquisition Policy
EBS	Electronic Business System
ECP	Engineering Change Proposal
EOQ	economic order quantity
FAR	Federal Acquisition Regulation
FSG	federal supply group
GAO	Government Accountability Office
LTC	long-term contract
MIS	Materiel Information Systems
MOQ	minimum order quantity
NIIN	National Item Identification Number
OSD	Office of the Secretary of Defense
PLT	production lead time
PPIRS	Past Performance Information Retrieval System
PR	purchase request
SSA	strategic supplier alliance

CHAPTER ONE

The Need for Increased Supply Chain Agility

The Long-Standing Inventory Challenge

In the Department of Defense (DoD), spanning the Defense Logistics Agency (DLA) and all of the military services, there has been significant pressure and emphasis on reducing inventory, as exemplified by a series of Government Accountability Office (GAO) reports and the DoD Comprehensive Inventory Management Improvement Plan.[1] One of the consequences of this pressure is that much attention has been placed on identifying inventory that can be disposed without affecting operations—that is, excess inventory. However, as will be discussed in this report, disposing of excess inventory results in only relatively small savings produced by a reduction in operating costs associated with storage. The larger cost associated with excess inventory is the sunk cost of the investment, which is not recovered through disposal. Without gaining and acting on an understanding of the causes of excess inventory, it is likely to build up once again, continually regenerating. Thus, the larger savings opportunity associated with excess inventory comes from changing processes so that less excess develops in the future. (The only supply chain that could achieve zero excess inventory is one that only produces in response to each order, with no customer cancellations or returns.)

This report focuses on this inventory challenge for DLA. To determine how much lower DLA's inventory could be and how it should change its inventory management

[1] See the following reports from the U.S Government Accountability Office: *Defense Inventory: Actions Underway to Implement Improvement Plan, but Steps Needed to Enhance Efforts*, GAO-12-493, Washington, D.C., May 3, 2012; *DOD's 2010 Comprehensive Inventory Management Improvement Plan Addressed Statutory Requirements, but Faces Implementation Challenges*, GAO-11-240R, Washington, D.C., January 7, 2011; *Defense Inventory: Defense Logistics Agency Needs to Expand on Efforts to More Effectively Manage Spare Parts*, GAO-10-469, Washington, D.C., May 11, 2010; *Defense Inventory: Army Needs to Evaluate Impact of Recent Actions to Improve Demand Forecasts for Spare Parts*, GAO-09-199, Washington, D.C., January 12, 2009; *Defense Inventory: Management Actions Needed to Improve the Cost Efficiency of the Navy's Spare Parts Inventory*, GAO-09-103, Washington, D.C., December 12, 2008; *Defense Inventory: Opportunities Exist to Improve the Management of DOD's Acquisition Lead Times for Spare Parts*, GAO-07-281, Washington, D.C., March 2, 2007a; and *Defense Inventory: Opportunities Exist to Save Billions by Reducing Air Force's Unneeded Spare Parts Inventory*, GAO-07-232, Washington, D.C., April 27, 2007b. See also U.S. Department of Defense, *Comprehensive Inventory Management Improvement Plan*, Office of Assistant Secretary of Defense for Logistics and Materiel Readiness, October 2010.

policies and processes, it may be tempting to look at industry supply chains that often operate with much lower inventory relative to demand levels. But a closer look will show that not all aspects of these models—and the relative inventory levels—effectively transfer to the DLA context. DLA's role within DoD is to provide consumable material, generally unique to or dominated by military demand within the government, for all DoD activities to accomplish their missions. The requirement to support the services' warfighting capabilities and needs means that DLA will often have to maintain inventories of items for which demand is highly variable even when the trend is stable, items for which demand could dramatically increase on short notice, and items for which the projected demand might never materialize. Private sector firms might consider stocking such items unprofitable and would not commit to doing so without a guarantee that the items would be sold, or at least without agreements with customers that could be counted on to produce an adequate financial return. DLA, with its mission to support the warfighter in the face of uncertain requirements, may not have this option.

A New Path for Further Improvement: Lead Time and Agility

Given this inventory management environment, what can DLA do then to reduce inventory—or does it even have the opportunity to do so? Much effort has been placed by DLA on improving methods for forecasting demand, particularly by using alternative mathematical models augmented with collaborative demand planning with customers. These efforts have resulted in continual refinements in the forecasting methods used with DLA's enterprise resource planning system. As is recognized in the academic literature and in the private sector, though, there is a limit on how good forecasts can ever be in the face of the long lead times involved in the bid and purchase process within DLA and in the manufacturing process at the supplier. Thus, further significant improvements in forecasting and inventory management will likely require reducing lead times throughout the entire supply chain that DLA manages. Such efforts to reduce lead time and other aspects of supply chain agility that we will define and discuss later reflect industry best practices and trends that should be emulated. Because improving DLA supply chain agility will require a DoD supply chain enterprisewide focus, in this report, we will walk through the supply chain for DLA-managed stocked material (excluding fuel), from the customers throughout DoD to DLA supply planning, to DLA acquisition, and to the supplier.[2] Each has a role to play in improving supply chain agility.

---

[2] DLA manages fuel and lubricants, subsistence (e.g., food), medical supplies, clothing and personal equipment, tools, construction material, and consumable spare parts for DoD. Most medical supplies, clothing, food, compressed gas, and packaged petroleum, oil, and lubricants are managed through prime vendor and other outsourcing arrangements, with material delivered directly to DoD customers by the suppliers rather than being

To start this discussion, it is helpful to briefly review the purpose of holding inventory—how it creates value—and why lead times are so important in inventory management. Buying inventory serves two purposes. The most important is having material on hand to serve customers. In DoD's case, this means

- having material on hand to support the services' maintenance depots by enabling production to proceed as planned
- providing material to units in the field to repair not-mission-capable equipment quickly enough to maintain sufficient equipment readiness for operations and to enable personnel readiness
- providing material to all other DoD activities to enable them to perform their jobs uninterrupted.[3]

The second reason is for efficiency, when buying in batches reduces procurement workload or materiel prices through quantity discounts enabled by more-efficient production.

To buy inventory and to support customers effectively by having inventory in stock when orders come in, DLA has to anticipate or forecast demand. The longer it takes to get something from a supplier—the *lead time*, including DLA's own internal processes to develop and place the order—the longer the horizon over which it must project likely demand. And the larger the order size relative to demand, the longer the horizon over which DLA has to project demand. The longer the forecast horizon, the greater the uncertainty in demand from both inherent variability in demand (e.g., from stochastic failure processes that produce variability around a mean time to failure) and trend shifts from changes in operational and equipping plans, including end item improvements (the longer the horizon, the greater the likelihood that an exogenous event will occur or a DoD decision will be made that affects an item's demand within the forecast horizon). These factors make forecasting difficult and prone to error.

Long lead times have several effects. First, they increase lead-time demand variability, driving up safety stock requirements to meet desired service levels. *Safety stock* is additional inventory purchased to account for a "known" level of variability in lead-time demand, which is estimated, enabling customers to continue to be satisfied until more materiel arrives in situations in which the lead-time demand is greater than the expected level. Second, even with higher safety stock, there will be times it will be difficult to support customers well because safety stock is not designed to account for unexpected trend shifts. If demand suddenly spikes, shifts much higher, or materializes

---

stocked in DLA distribution centers. Some items in other categories also have customer direct contracts. The scope of this report is on those items stocked by DLA.

[3] In all three cases, this is done either through relatively rapid replenishment of retail stocks, which tend to be relatively thin, that provide materiel to local customers immediately upon demand or in direct response to customer needs.

for an item that is not stocked, with long lead times, the stockout or backorder periods can be lengthy. Third, if demand for an item declines due to a trend shift—including totally drying up due to a change in operational use or equipment design—the inventory of the item will become excess. Excess inventory that is ultimately disposed represents a cost of business—a cost of supporting customers that has to be borne. As will be discussed in more depth later, this is the key, tangible component of inventory holding cost for DLA.

While customers are not always happy—because the DoD goal, as with any business, is not 100-percent satisfaction—DLA has made strides in achieving customer support goals. For example, Figure 1.1 shows that materiel availability has gotten close to or achieved the targeted 90-percent level promised to the services across the hardware supply chains. This is as agreed to or even higher than expected by the services in some cases.

The Inventory Holding Cost of Not Being Agile

Achieving this high level of customer support has come at a significant cost in inventory, manifested to a large degree in excess inventory that eventually leads to disposals. As shown in Figure 1.2, from 2005 through 2013, DLA disposed of more than $1 billion per year, albeit with significant variability when "the garage has been periodically cleaned out."[4] This represents prior annual spending or materiel cost that was ultimately was not necessary; in business, this would be equivalent to an annual $1 billion write-off and charge to the income statement. To put this in context, this is 14 percent of the annual DLA sales from inventory over this period.[5] However, there is no clear industry benchmark for DLA to judge whether this is too high or what it should be. To establish such a benchmark would require identifying an industry sector with the same set of supply chain characteristics (e.g., lead times, demand variability and uncertainty,

---

[4] This became inflated somewhat beyond normal inventory excess generation because, starting in late 2012, DLA began reevaluating *contingency retention levels*, or the amount of inventory above current requirements kept on hand for possible contingency operations. This has led to reduced contingency levels, resulting in pulling ahead some disposals to create the large 2013 spike. This is part of a broader effort across DoD to reevaluate retention levels and dispose of retention stock. These DLA disposals compose only part of a DoD focus on reducing inventory, which also includes large disposals by the services. DoD policy authorizes two types of retention stock. *Economic retention stock* is on-hand stock above requirements, based on forecasts and inventory determination procedures and policies, that is deemed more economical to retain than to dispose of. This is based on the probability of future demand, the cost of storage, and the cost of having to repurchase the item later. *Contingency retention stock* beyond the economic retention stock level is authorized if the level of stock is determined to be needed for specific contingency operations with the justification documented. U.S. Department of Defense, *DoD Supply Chain Materiel Management Procedures: Materiel Returns, Retention, and Disposition*, DoD Manual 4140.01, Vol. 6, February 2014b.

[5] This is based on the dollar value of issues of DLA-owned and managed stock from DLA distribution centers. See Appendix B for a complete discussion of DLA inventory holding costs and the contribution of these disposals.

Figure 1.1
DLA Hardware Supply Chain Materiel Availability Trends Versus the Goal

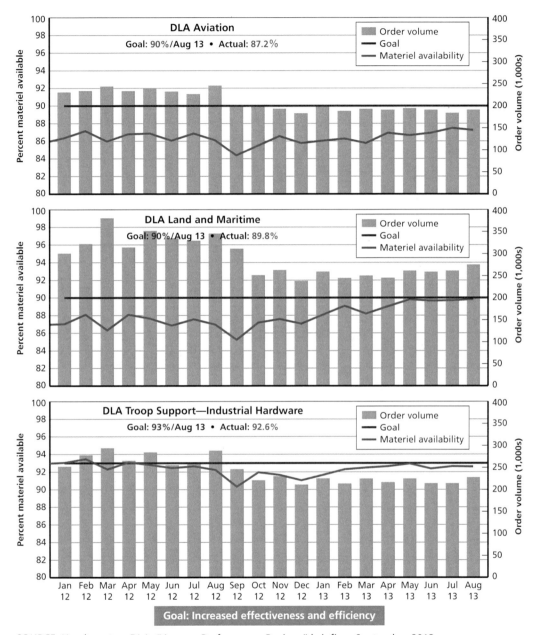

SOURCE: Headquarters DLA, "Agency Performance Review," briefing, September 2013.
NOTE: This figure shows the materiel availability trends for the three DLA organizations, called supply chains, for which materiel availability is measured.
RAND RR822-1.1

Figure 1.2
DLA Serviceable Item Disposals, by Calendar Year

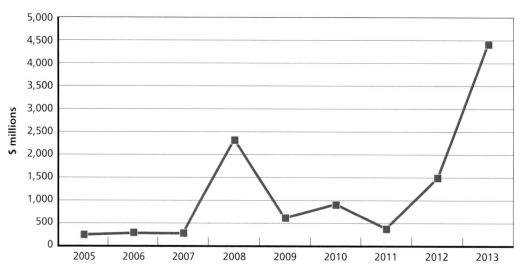

RAND *RR822-1.2*

and commodity or item mix) as DLA, or, more precisely, a basket of industries with the same supply chain characteristics as the different DLA product segments. Differences in these factors lead to dramatically different levels of excess inventory and overall inventory management efficiency for different industry segments.

So rather than comparing DLA disposals with a benchmark level, we provide these figures (total disposals and disposals as a percentage of sales) to help make the cost of managing inventory more transparent and give a baseline from which process improvements can be used to measure success. This cost is not typically tracked within DoD and DLA, whether for updating inventory holding cost estimates or tracking progress in inventory management. In DoD, these inventory write-offs are a cost of doing business that DLA must charge to its customers through its cost recovery rate or surcharge. The price that its customers pay—the services and others—is the latest acquisition cost of an item plus the cost of providing materiel and holding inventory, to include the cost of having to buy some inventory that is not later consumed in order to ensure that desired service levels are met.[6] Effectively managing all of the cost elements of the surcharge requires a detailed understanding of each of the components.

---

[6]   DLA and other DoD supply organizations recover their costs by charging prices based on the acquisition cost of goods sold and a surcharge to cover their operating costs. Inventory that is purchased that becomes excess and is disposed of requires spending that is not recovered through a sale. This gap has to be made up through the surcharge applied to material that is sold. In short, due to forecast error, DLA and other DoD supply chain organizations will always have to buy more than they sell over the long run, with this difference being made up in the cost recovery rate or surcharge (or it can be made up through periodic direct appropriations for the purchase of

The Customer Service and Excess Inventory Risk Trade-Off

As implied, excess inventory is not unique to DLA or DoD. Every business has unsold inventory; the difference is that most businesses are able to first try to gain as much value from it as possible through sales or offering the product through lower-cost outlets. As long as a business employing these alternatives can still sell inventory at prices creating revenue greater than the operational cost of having a sale or selling through an alternative outlet (i.e., the marginal cost of using the alternative sales channel or conducting the sale is less than the revenue produced), it is worth it to reduce the price or employ alternative sales channels because the initial investment in the inventory is a sunk cost. Once the inventory is purchased, it no longer becomes necessary to recover the total cost to purchase the inventory plus make a profit. While businesses would prefer to reduce the amount of inventory they write off or the amount of discounts they offer, they focus on maximizing profit, which also depends on having items to sell when a customer wants to buy them. The dynamics of each product segment and a company's competitive positioning affect where a company wants to be in terms of the trade-off between the risk of reduced revenue from a lost sale from not having an item in stock when a customer wants to buy it and the risk of increased cost from buying inventory it cannot sell.[7]

Determining the ideal trade-off—not too little or too much inventory, but just right—for profit maximization for DLA is more difficult because the "cost" of a backorder may not be directly lost dollar sales and profit but rather lower readiness, a depot maintenance production disruption or workaround that impedes efficiency, or simply customer dissatisfaction, depending on the reason for the requisition. Or even when there is a financial impact, such as a depot maintenance disruption, the value of the impact is not known and is difficult to estimate. This leads to somewhat arbitrary but reasonable service-level goals, generally codified in performance-based agreements reflecting negotiations between the services and DLA, historical levels, and reactions to customer complaints. With regard to customer complaints, DLA can be thought of as also reacting to competing pressures and making the trade-off between minimizing total complaints from the services for poor customer support and minimizing pressure from the Office of the Secretary of Defense (OSD) and external stakeholders (such as Congress) for holding and disposing of too much inventory.[8]

---

inventory). Having material on hand to immediately satisfy customer needs is a service that the supply organizations provide, and this is an unavoidable cost of this service.

[7] In contrast, a company is always interested in reducing inventory if it can do so without affecting sales and if it can do so at no or less cost than the cost of holding the inventory (i.e., if the costs of activities or process changes needed to reduce inventory, such as new production methods, are less than the costs of holding inventory). This would stem from process improvement.

[8] Through dozens of projects involving the services and DLA and extensive engagement with personnel at DLA, the authors have observed what they believe to be DLA reactions to both of these competing pressures.

So, assuming DLA's customer service goals will remain constant, to what extent and how can DLA reduce inventory and, more importantly, the annual $1 billion in spending on items later disposed of while still meeting these goals? Prior RAND research found that lead time is the strongest management factor associated with the buildup of DLA inventory excess (as measured through inventory turns and large increases in on-hand inventory relative to demand, with each such instance increasing the chances of long-term excess) and customer support effectiveness (as measured through materiel availability and the length of stockout periods). That is, differences in lead times among unique items identified by their National Item Identification Numbers (NIINs) were found to be strongly associated with substantial differences in relative inventory levels and customer support effectiveness. Notably, the effects of order quantities could not be analyzed because DLA practices have limited the order quantity differences among NIINs.[9]

Demonstrating How Improved Supply Chain Agility Could Help DLA

These effects are easiest to demonstrate and explain through examples. Figure 1.3, as with the three following figures, illustrates this with one example DLA-managed item. It shows monthly replenishments from the supplier, demands (month orders were placed), and on-hand inventory. Through September 2008, demands ran about 20,000 per month, with substantial variability and some occasional higher spikes, but no apparent upward or downward trend. The early demand spikes seen in 2006 and 2007 led to inventory being drained to zero. It then recovered, with inventory remaining between just barely enough on hand to about five months of supply on hand from mid- to late 2008, supporting customers well without building significant excess. But then demand suddenly dropped in October 2008 and stayed low, reflecting a trend shift to a new, much lower level. However, given the item's lead time and the time it takes for the statistical forecast to adjust when based on historical information (quick forecast adjustments combined with long lead times would lead to highly volatile supply plans that overreacted to short-term demand variability around a mean), deliveries stemming from orders to replenish at the previous 20,000 monthly demand rate were still incoming through January 2010, building up substantial inventory excess.

Some inventory was drained during a temporary demand surge in mid-2010, but this left substantial excess in comparison with ongoing demand. In this case, the potential excess inventory was largely divested through large military assistance program issues, so some value was gained in either building partner capability or goodwill, or both. In other cases, this is not an option or not deemed valuable, with such excess

---

[9] E. Peltz and M. Robbins, with G. McGovern, *Integrating the Department of Defense Supply Chain*, Santa Monica, Calif.: RAND Corporation, TR-1274-OSD, 2012. The research did not include studying how or to what degree DLA could improve lead times.

Figure 1.3
Replenishments, Demand, and On-Hand Inventory for Example Item A (unit price $6)

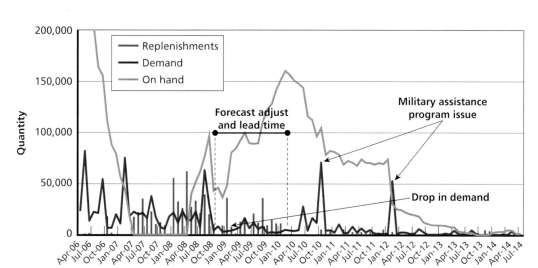

SOURCES: On hand: Quantity by Owner file; Demand: DLA Materiel Information Systems (MIS) Issues file; Replenishments: DLA MIS Receipts file.
RAND *RR822-1.3*

resulting in disposals. Thus, this is still an instructive case for understanding the value of becoming more agile. Being able to react faster to the sudden decline in demand would have prevented this inventory buildup. The buildup also would have been prevented if the customer organization knew that demand was going to decline; the using service or cognizant program manager relayed this information, which provided a lead time in advance of the decline; and then DLA acted upon the information. Even if it was not a full lead time in advance, such information could have been used to reduce the buildup.

Figure 1.4 shows a somewhat similar situation, with the difference being a temporary but sustained upward demand shift followed by a return to low demand. The increased demand lasted long enough to appear to be the new normal, lacking information from customers to the contrary. This led to many years of supply being on hand and then demand eventually going close to zero, leading to a large disposal in late 2013 with substantial inventory still remaining.

The length of the lead time, however, does not just affect the risk of building up excess inventory due to demand shifts. It also changes the nature of customer support effectiveness in the face of demand shifts and spikes. The example in Figure 1.5 illustrates this. A demand spike in October 2007 led to depletion of the entire stock of inventory (the full depletion was temporarily delayed until early 2008 by rationing the release of on-hand material to fill only the highest-priority orders). In this case, due to the long lead time, it took over a year—until January 2009—for the replenishment

Figure 1.4
Replenishments, Demand, and On-Hand Inventory for Example Item B (unit price $2,700)

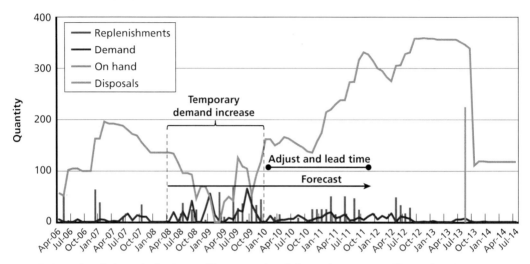

SOURCES: On hand: Quantity by Owner file; Demand and disposals: MIS Issues file; Replenishments: MIS Receipts file.

RAND *RR822-1.4*

Figure 1.5
Replenishments, Demand, and On-Hand Inventory for Example Item C (unit price $78)

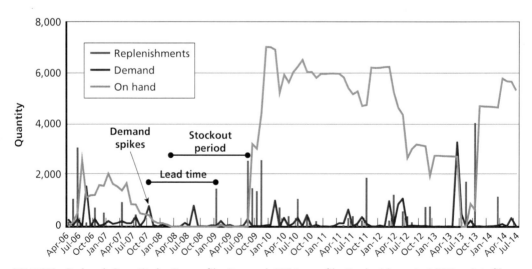

SOURCES: On hand: Quantity by Owner file; Demand: MIS Issues file; Replenishments: MIS Receipts file.

RAND *RR822-1.5*

order placed in response to the demand spike to be delivered. In the meantime, another demand spike had occurred in mid-2008, so the first replenishment enabled some outstanding backorders to be filled, but the on-hand level remained at zero until the next replenishment in August 2009, which was placed in response to the mid-2008 demand spike, creating an extended stockout period of more than one year. So, this graph shows another aspect of customer support besides an aggregate measure of the service level or materiel availability. What the aggregate numbers mask is that for a specific item, the shelves can be bare for an extended period of time in the face of demand spikes combined with long lead times. This can have significant concentrated effects, such as preventing depot production of a depot-level repairable for a long period of time. In this case, keeping a higher level of inventory eliminated stockouts or kept them limited to a short period of time in the face of subsequent demand spikes in 2012 and 2013.

We now turn to how ordering in large batches can further increase the risk of being in an excess inventory position, leading to disposals. (And, of course, the true cost or risk is spending money on material for which zero value is gained.) Ordering in large quantities increases this risk in two ways. The first is that if demand drops unexpectedly, more of the inventory will become excess. The second is that it often takes longer for the supplier to produce and deliver the full order, increasing the lead time and thus the forecasting horizon, thereby increasing the chance of a demand decrease before the full order comes in and the inventory is consumed. The next example illustrates these effects.

Added to the graph in Figure 1.6 are three procurement actions, with the gray dot indicating the order placement or contract date for each, the pink dot indicating the first delivery associated with the order, and the brown dot indicating the completion of deliveries associated with the order. First, because the orders were with the same supplier and overlapped, the start of deliveries for the second and third orders got progressively further from the contract date, producing progressively longer lead times for the start of deliveries. In effect, these three orders became like one very large order, with lead time extended by the combined order size. As a result, the third order of 80,000 units had a total time from the start of deliveries until delivery completion that was twice as long as for each of the two 40,000-unit orders. In this case, order size was directly related to lead time.

After the orders were placed, the demand was much lower than the forecast, leading to substantial inventory excess, with the beginning of disposals from inventory being seen in May and December 2013. Two things happened to cause this. First, there had been a plan to recapitalize a significant number of the vehicle for which this item is a component, and the service planning organization notified DLA of the plan. That plan was canceled, but DLA had leaned forward and placed orders of this item in sufficient quantity to support the complete recapitalization program. So, the intent was to serve the customer well. But not all of the material had to be ordered up front because the production to recapitalize the vehicle would have been conducted over a

Figure 1.6
Replenishments, Demand, and On-Hand Inventory for Example Item D (unit price $45)

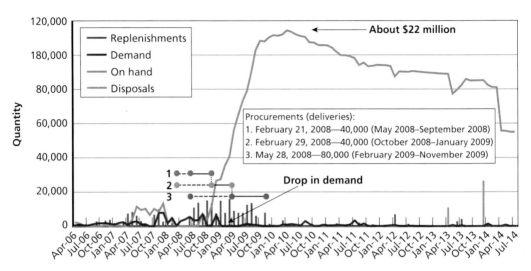

SOURCES: On hand: Quantity by Owner file; Demand and disposals: MIS Issues file; Replenishments: MIS Receipts file.

RAND *RR822-1.6*

long period of time. Instead, the item could have been ordered gradually over time, with deliveries scheduled as needed in concert with the production schedule. Second, at almost the same time that this recapitalization plan was canceled, the service that primarily used this vehicle started using a new and improved version of this item on the vehicles in the field for improved durability in order to reduce maintenance costs and improve readiness. This led to dramatically lower demand for the example item, reflecting the demand drop that started in January 2009.

Defining Supply Chain Agility and What It Means for DLA

What can be done to reduce the risk of inventory excess and thus reduce disposals and material costs? And at the same time, what can be done to reduce the length of stock-outs from surprise demand increases to qualitatively improve customer support? The answer is improving supply chain agility, or being able to better keep supply aligned with demand,[10] in order to serve customers as well as or better than before, but at lower cost—to potentially include having less inventory on hand or on order. More generally, we define improving supply chain agility as improving the responsiveness and

---

[10] This is similar to the definition offered in X. Li, C. Chung, T. J. Goldsby, and C. W. Holsapple, "A Unified Model of Supply Chain Agility: The Work-Design Perspective," *International Journal of Logistics Management*, Vol. 19, No. 3, 2008.

efficiency with which customers with changing needs are served by increasing supply chain alertness and the rapidity of supply chain response to changes in demand and supply. This can include changes specific to individual items and those from broader changes in the environment that affect item segments or all or much of the item population.[11] Research has shown that "the more dynamic the customer needs and expectations, the greater the necessity for higher levels of [supply chain agility]."[12] DLA clearly faces very dynamic demand that presents a challenge for efficient and effective inventory management. The dynamism of supply appears to be much less, with major problems from supply-side volatility not apparent or documented in prior research. Thus, in this report we focus on the agility implications of being able to better respond to volatile, uncertain demands, particularly at the item level. This has three important components for DLA.

The first component for improving agility is improved advance communication of customer plans that produce demand spikes, cutoffs, or dramatic trend shifts for an item or a group of related items—along with DLA action taken in response to this information. At the macro level, DLA is well informed of major changes in the defense environment and operations, and DLA and DoD take steps to position DLA to respond to aggregate changes in demand, such as through budget adjustments. However, at the micro level, anticipated demand changes for specific items that stem from general environmental shifts or from smaller changes, such as product design changes, are not shared as well, with respect to both general communication and the transfer of sufficiently detailed data for implementation planning. Improving on this dimension requires that DLA and its customers be alert to changes, that DLA has access to the needed data to respond, and that DLA quickly decides to take action.[13] One might say that this alone should be the solution, but sometimes the changes in demand are due to surprises that the customers face as well, such as an unexpected operational requirement, an unexpected funding cut to a program, or a vehicle safety problem that newly appears. Or the lead time for the customer change of plan may be less than the lead time to procure the item. So, even if DLA and its customers could perfect collaborative planning, this alone would not eliminate disposal or customer service risk.

---

[11] For a useful review of supply chain agility definitions, see D. M. Gligor and M. C. Holcomb, "Understanding the Role of Logistics Capabilities in Achieving Supply Chain Agility: A Systematic Literature Review," *Supply Chain Management: An International Journal*, Vol. 17, No. 4, 2012b.

[12] D. M. Gligor and M. C Holcomb, "Antecedents and Consequences of Supply Chain Agility: Establishing the Link to Firm Performance," *Journal of Business Logistics*, Vol. 33, No. 4, December 2012a, pp. 295–308.

[13] Recent work on conceptualizing supply chain agility concludes that it has fives dimensions: alertness, accessibility (of data), decisiveness, swiftness, and flexibility. These dimensions are referred to in the discussions of the three components of supply chain agility for DLA laid out here. See D. M. Gligor, M. C. Holcomb, and T. P. Stank, "A Multidisciplinary Approach to Supply Chain Agility: Conceptualization and Scale Development," *Journal of Business Logistics*, Vol. 34, June 2013.

The second component for improving agility is finding ways to shorten the lead times it takes for DLA to procure materiel and thus more quickly respond to changes in demand, thereby reducing the risk of under- and oversupply. This requires quickly implementing changes to plans (including placing orders) and having the flexibility in the supply base to enable rapid delivery.[14] As will be discussed in more depth later, this is not about paying more to get faster deliveries but finding process improvements to do so—or only paying more when the extra cost has a positive expected value when simultaneously considering the reduced inventory disposal risk that results from shorter lead times. Hand in hand with shortening lead times is streamlining and reducing the workload burden associated with procurement. This reduced workload frees up procurement labor and resources. Supply responsiveness is also affected by the level of visibility into supply base capabilities, alertness to changes in the supply base, and responsiveness to these changes to ensure that supply flexibility is available when needed.[15]

The third component for improving agility is ordering smaller, more-frequent quantities relative to demand, which decreases the risk of excess from deliveries continuing despite demand decreases, in line with maintaining flexibility. Again, this should not be done in isolation in a way that increases total costs, but rather this should be done using optimal order quantities from a total cost perspective, including item prices. Further, acquisition process improvements can reduce the optimal sizes of orders and total costs. In concert, shorter lead times and smaller order quantities mitigate the inventory excess and shortfall consequences of not getting available customer plans or customers facing demand surprises themselves. By reducing the unnecessary purchases that lead to inventory excess, improved agility frees up resources that can be directed to other needs.

Project and Report Overview

As described earlier, prior RAND research identified inventory purchased but not ultimately needed—reflected in inventory that becomes excess and is disposed of—as the biggest driver of DLA inventory holding costs and determined that long lead times have been the biggest contributor to this cost of buying unneeded materiel.[16] Based on this research, the Assistant Secretary of Defense for Logistics and Materiel Readiness commissioned a RAND study to work with DLA to identify paths to improving its supply chain agility encompassing the three components identified: lead times, order quantities, and customer collaboration. The assistant secretary requested that we provide general recommendations akin to commanders' intent, with DLA then having

---

[14] Gligor, Holcomb, and Stank, 2013.

[15] Gligor, Holcomb, and Stank, 2013.

[16] Peltz, Robbins, and McGovern, 2012.

the responsibility to develop them in sufficient detail for implementation. This takes advantage of DLA's internal process knowledge and preserves flexibility to adapt the recommendations as further learning occurs. The scope includes all material that DLA stocks, of which repair parts are the dominant portion. The research was accomplished through a multimethod approach consisting of a literature review, extensive interviews and discussions with DLA personnel and suppliers, discussions with non-DoD suppliers, a policy review, and quantitative data analyses where applicable and where the requisite data were available. This report highlights key findings and the resulting recommendations. More-detailed findings, along with additional background information, to support the recommendations are in technical appendixes to this report.

Chapter Two discusses the central recommendation of this project, which is to increase an enterprisewide emphasis on supply chain agility. The remaining chapters focus on different points in the supply chain, beginning with customers, moving through DLA, and ending with suppliers. Chapter Three presents findings and recommendations regarding collaboration with DLA's customers, and Chapter Four presents the same for determining coverage duration and order quantities. Chapter Five presents findings and recommendations for acquisition processes, and Chapter Six addresses supplier management and integration. The report concludes with a summary of recommendations in Chapter Seven.

CHAPTER TWO

Increased Enterprisewide Emphasis on Supply Chain Agility

DLA's Key Outcomes

For DLA, two outcomes ultimately matter: how well it serves customers and how much it costs to do so. At a high level, serving customers is measured through materiel availability or the overall service level. Costs consist of internal operating costs and the cost of material. The cost of material for stocked items consists of the cost of the material—the purchase price paid by DLA—and the costs associated with holding inventory. Those latter costs vary with inventory turns, which measure the efficiency of inventory management for a given level of customer service. From this vantage point, there are three key outcomes for those in DLA responsible for managing stocked items: materiel availability, inventory turns, and purchase price.

How Processes Interact to Produce the Key Outcomes

Figure 2.1 shows the primary factors and processes that affect these three outcomes. Note that all of these factors ultimately affect at least two of the outcomes, and sometimes are related to all three. So, there are interactions among the outcomes. In some cases, changes in one factor lead to changes in a common direction (in terms of being positive or negative), and in other cases, there are trade-offs involved. The higher the service level target, the higher materiel availability becomes (positive) but the lower inventory turns becomes (higher costs, which is negative). Improving demand forecast accuracy has positive effects on both inventory turns and materiel availability, as does improving both of its main components, forecast methods and customer information flow. Similarly, lead-time improvements are positive for both turns and availability. While a customer may sometimes have to pay more for reduced lead time, this does not necessarily have to be the case. For example, improved supplier selection, collaboration, and joint process improvement can produce shorter lead times at potentially similar or even lower prices. Note, too, that improving lead times improves forecast accuracy, magnifying the effect on turns and materiel availability. Decreasing order quantities involves a trade-off, producing higher turns but lower availability through

Figure 2.1
Key Supply Chain Outcomes and the Factors That Drive Them

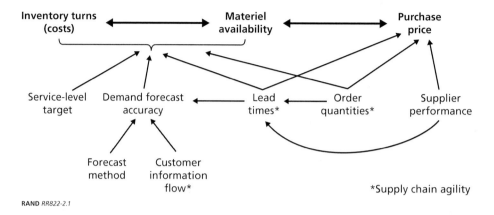

RAND RR822-2.1

direct effects. However, decreasing order quantities can also reduce lead times, which can produce higher turns in tandem with higher availability. And there may be a direct item price or cost trade-off involved in decreasing order quantities, as the higher turns (and lower inventory cost) may come with a higher purchase price (but like lead times, there is not always a direct relationship between order quantities and price). Further, lower order quantities could increase operating costs through increases in transactions (not shown). Conversely, lower order quantities could also increase materiel availability in some cases because less working capital is tied up for each item, preserving financial flexibility to respond to demand changes, which could come into play if and when financial limits on placing orders are hit.[1] Supplier performance improvement should be positive for both purchase price and lead times, creating a potential win-win-win among the three outcomes. In sum, the outcome metrics should not be evaluated in isolation, and doing so can be counterproductive for management decisionmaking because of these complex trade-offs.

Figure 2.1 also shows how important supply chain agility is to outcomes, both cost and performance, which is consistent with the introduction in Chapter One and the motivation for this project. As shown by the asterisks in the figure, three of the key factors represent our working definition of how supply chain agility can improve responsiveness to demand changes. Notably, two of the three do not involve trade-offs between turns and materiel availability, with improvements in these two factors being positive for both outcomes. Although it is possible that better lead time may involve a trade-off with an item's purchase price, this is only the case sometimes. Still, it is important to remain cognizant of this potential trade-off, and those trade-offs involved with changes in order quantities.

---

[1] DLA can place orders with suppliers up to an annual limit constrained by its obligation authority budget. While this can be increased in an environment producing an increase in aggregate demand, it is possible that this may not happen or may not do so in time to place orders when needed to maximize customer support.

Functions and Organizations That Affect the Processes and Outcomes

Table 2.1 shows the major organizations (customers, DLA, and suppliers) and DLA supply chain functions (cutting across customer operations, supplier operations, and procurement and acquisition organizations) involved in producing the three supply chain outcomes, along with which functions affect the three processes of supply chain agility.[2] Because all of the functions and organizations listed affect one or more components of supply chain agility, they all affect the outcomes (and suppliers, buyers, and postaward management influence price changes through supplier performance and relationship management). Two things should jump out. Every outcome is influenced

Table 2.1
Interaction of Supply Chain Functions, Outcomes, and Agility Processes

Function(s)	Customers	DLA							Suppliers
	Engineering Services, Demand	Weapon System Program Manager[a]	Customer Account Specialist[b]	Demand Planner	Supply Planner	Buyer	Post-award Manage-ment	Bid Development, Order Fulfillment
Supply Chain Outcomes									
Materiel availability	X	X	X	X	X	X	X		X
Inventory turns	X	X	X	X	X	X			X
Price changes					X	X			X
Supply Chain Agility Processes									
Lead time	X					X	X		X
Order quantity					X	X			X
Customer information flow, or forecasting	X	X	X	X					

[a] A weapon system program manager is a DLA employee on site with major weapon systems to provide service and facilitate information flow.

[b] A customer account specialist represents DLA's customer-facing personnel who provide a central point of contact to DLA's customers for service. Customers are the services and other organizations. Within the services, Engineering Services can affect lead times through its roles in providing approvals (e.g., first article testing) and decisions about item upgrades and substitutions. In addition, any organization that creates plans that affect demand has a role in customer information flow and forecasting.

---

[2]   The DLA supply chains are organizational entities encompassing all of the functions related to providing material for a product segment. DLA's supply chains include Aviation, Land and Maritime, Troop Support, and Energy, and the scope of this report is limited to the first three.

by multiple supply chain functions and organizations—including all for materiel availability and almost all for inventory turns. Similarly, all components of supply chain agility are affected by multiple functions and organizations.

This means that no single person or organization can be held accountable for any outcome and process performance metric. This is consistent with the commercial sector, where integrating purchasing and logistics remains a persistent challenge to supply chain efficiency.[3] At DLA, the first place in the structure where procurement, customer operations, and supplier operations fully come together is at the DLA supply chain commander level, encompassing everything from weapon system program managers to supplier management. However, DLA Headquarters' policies and oversight also affect these outcomes and processes, moving the only place at which these outcomes come together up to the DLA director level. The services (including their engineering services divisions) are the customers, and they also have a role to play in their own support because they affect lead time and forecasting. Thus, the only place these outcomes and factors all come together, encompassing suppliers and their performance, from a management perspective is at the DoD level. But DLA and the services can also jointly manage these outcomes through their various joint management meetings (e.g., periodic DLA service meetings) and other processes.

Making Agility Part of the Enterprise's DNA

The central recommendation of this project, which will be expanded upon later, is to dramatically increase the focus on supply chain agility for DLA-managed material across the DoD supply chain enterprise, which starts with top leadership attention and requires threading such an emphasis through all processes. For DLA to fully embrace and achieve its supply chain agility potential, agility has to be part of the mind-set for constant awareness and consideration in all process decisionmaking—and this has to be extended to collaboration with customers and suppliers through DLA's efforts and with OSD oversight. Supply chain agility is driven by many elements that need to be individually acted on but should also be viewed holistically, bringing these elements together for a total picture of the level of agility and how it is changing over time. The processes that affect supply chain agility for DLA cross organizational boundaries within and outside of DLA and within and outside of DoD, and the increased emphasis on supply chain agility similarly needs to include all of these organizations and how DLA works with them and across its own organizational boundaries. Such an enterprisewide focus necessarily requires specific actions for multiple processes and at multiple levels, which we describe below. Some of these efforts are already under way, especially within DLA.

---

[3]   T. Stank, J. P. Dittman, C. Autry, K. Petersen, M. Burnette, and D. Pellathy, "Bending the Chain: The Surprising Challenge of Integrating Purchasing and Logistics," Knoxville, Tenn.: University of Tennessee Global Supply Chain Institute, Spring 2014.

Because supply chain agility crosses organizational boundaries, an increased emphasis requires senior management–level involvement, starting with the Assistant Secretary of Defense for Logistics and Materiel Readiness to ensure that the services and DLA work in concert and with the director of DLA, cascading through all levels within DLA. Everyone has to know agility is important. This comes through several forms: ensuring process integration of agility considerations, employing the right metrics and holding personnel accountable for them, appropriately asking relevant questions in briefings and reviews, and training to ensure that all stakeholders understand the importance of agility and how they affect it. Multiple points on the supply chain, multiple functional fields, and multiple organizations jointly determine both the enterprise's three central outcomes with respect to stocked items (materiel availability, purchase price, and inventory turns—a proxy for cost) and supply chain agility capabilities that affect these outcomes (lead times, order quantities, and the flow and use of customer demand information). Management-level action can cross these boundaries and increase the emphasis in multiple ways.

Specifically, management can drive the consideration of lead times, order quantities, and potential changes in demand throughout all supply chain processes within DLA, as it is doing with the Time to Award team, for example. DLA management also has to ensure that the services understand their roles in DLA's supply chain agility and develop means for ensuring that they focus on this in their process design and improvement efforts. For the services, two sets of processes are directly relevant. Engineering Services processes (e.g., for first-article testing) affect administrative lead time (ALT). And information about planned changes in an item's demand, such as upgrades or changes to NIINs, changes in use of a NIIN, and schedules for repair, are critical for demand planning.

Within DLA, many positions and organizations affect supply chain agility. Processes for seeking and collecting information about changes in demand are spread across demand planners and all customer-facing roles, which demands integration across organizations even at this process level. Lead time is affected on one end of the supply chain by customers through their engineering activities, which can have a significant effect through necessary approval processes, such as first-article testing. Lead time is also affected by all acquisition functions involved in selecting and managing suppliers, with suppliers being the final key organization affecting lead times. Order quantities are managed and driven by a similar group as lead times, with the addition of supply planners.

Fundamental to reinforcing management emphasis is the use of metrics for oversight, accountability, and incentives, with incentives limited in this context to the competitive spirit of DoD personnel (e.g., seeing how their organizations compare with others at meetings or metrics reports), recognition through awards and at meetings, inclusion in and impact on performance reviews, and consideration in selecting personnel for promotions. Additionally, because outcomes and agility are affected by all organizations, achieving goals has to be a group endeavor with common metrics viewed together; they cannot be assigned to a lead organization. This requires putting emphasis on total enterprise performance rather than functional performance. In DLA's monthly Agency Performance Review (APR), there has long been a strong emphasis on materiel availability, with less emphasis on the other two outcome metrics described here. While broader outcome and agility process metrics are used in some standing functional meetings, all of the high-level outcome metrics should be emphasized to the same degree as materiel availability by showing them at APRs to ensure a balanced view and to recognize the trade-offs that can occur among them (the other metrics are computed but not necessarily used as primary metrics and shown at the APRs). This will avoid focusing too much on one metric at the expense of another, and it will communicate that they are all important to the director. Additionally, while inventory turns are computed, we recommend that the metric include all inventory in the computation, including retention stock. DLA should include award price change in the metrics shown at APRs and should modify it to be a comparison to a relevant weighting of producer price indexes.

We also recommend adding agility metrics to DLA's monthly Inventory Management Council report. The current report includes process metrics that can be augmented with additional focus on supply chain agility. These additional metrics are ALT, production lead time (PLT), and order quantity overrides. They would enable the Inventory Management Council to track lead times and compliance with optimal order quantities. Many overrides could also indicate a need to adjust order quantity computation parameters based on such constraints as procurement workload. Or they could be a sign of a process problem. We recommend extending the use of DLA's demand forecast accuracy to measuring the services' performance with regard to collaboration. Currently, this metric is only used within DLA to evaluate its own performance, but the services play a key role in the flow of information about changes in demand, and applying the metric to them as well captures this influence.

In addition to management emphasis and metrics, a third way in which management can increase the consideration given to agility is through training. It is difficult to ask people to focus on improving a capability when they do not have a clear view of how it will lead to improvement in overall outcomes or all of the effects they may have on it. The many personnel with whom we spoke were often unclear about supply chain agility, how it can affect all elements of value, and their influence on the different aspects of agility. We recommend introducing information on supply chain agility into the training material for DLA's customer-facing, planning, and contracting communities. Training material needs to include information on the effects that each community has on unneeded buys (and the resulting risk of disposal), on stockouts, and on customer support. It also needs to include guidance on how to minimize and balance the various risks in job tasks. This training needs to be accompanied by tools to evaluate the trade-offs, and such tools will be discussed later in this report.

CHAPTER THREE

Customer Processes

Improving the flow of information from customers to DLA is an essential component of improving DLA's ability to respond to changes in customer demand. Although the inherent uncertainty associated with military operations means that changes in demand are inevitable, not all changes in demand are entirely unpredictable. Some may be the result of decisions made by the services that can be better conveyed to DLA.

DLA has efforts with the services to conduct collaborative demand planning, especially with the depots, which conduct overhaul programs and repair reparable items. These industrial locations are high-volume customers but have demands that, to some extent, are planned in advance. While not perfect, processes are in place to share information and to improve the forecasts.

In this chapter, we describe another area where information can be shared: changes in the use of NIINs that result from engineering changes.[1] The choice of NIIN that is used for a system may change for a variety of reasons. It could be that a decision is made to replace the NIIN with one that is stronger or more durable. Or the entire subassembly or subsystem in which the NIIN is used is replaced with a new subassembly that does not use that NIIN. In cases of safety of use or safety of flight, these changes might be made on very short notice. In other cases, the changes could be the product of a long study and approval process. If DLA is not notified in a timely manner of these changes, it could end up with an oversupply of the old NIIN and an undersupply of the new NIIN, simultaneously creating excess inventory and a customer support shortfall (if the old item is not substitutable for the new one—and customers balk at using the old one). However, due to the long lead times associated with procuring parts, even when a service notifies DLA as soon as it knows about a change, the amount of notice might not be long enough to avoid ordering the old, wrong item within a lead time in advance of the cessation of use or the desired switch. This situation could be mitigated or avoided altogether with shorter lead times, which will be the focus of Chapters Five and Six. We interviewed personnel in two of the services, the Army and the Air Force, to learn how information about such changes is

---

[1] In this report, as is common practice in DoD, we use *NIIN* to mean both the identification number of the item and the item itself.

currently conveyed and how the process could be improved. Details about the interview methodology can be found in Appendix A.

Findings

Despite the fact that the services and DLA share the common objective of improving forecasts to reduce costs and improve materiel availability, they have different near-term driving goals, which can affect information-sharing. DLA's goals are to meet customer needs and minimize inventory excess. To pursue these dual goals, DLA needs to know about potential changes in the use of NIINs as soon as possible. The sooner that DLA knows about changes, the sooner it can cease to order the old NIIN, more completely use up its remaining supply, and begin to order the new NIIN. While the services ultimately benefit from DLA's ability to control costs by minimizing inventory buildup in this way, their more immediate goal is to ensure materiel availability from DLA. The effects of excess inventory on DLA's prices and thus the services' costs are much more indirect, and these effects are likely not even clear to most supply chain management personnel in the services. Plus, they are distributed across items and the services. The services thus aim to ensure that DLA avoids premature reductions in orders for an item so that there is no risk of it being phased out too early, especially in cases where the services decide not to make a change after all. To address this need, the services reported that they tend to share information relatively later in the process of changing a NIIN.

The other major finding is that information-sharing requires a great deal of manual intervention. Even though the services and DLA have automated systems, the different timing and item grouping of data systems means that information cannot flow easily between them, increasing the requirement for manual input. In addition, there is not an automated process for engineering changes to be automatically transferred into the DLA system; once a change is made, manual input is required.

In turn, manual input reduces the amount, frequency, and accuracy of information. For example, efforts to use estimates of flying hours to improve forecasts for service parts are hampered because, while DLA forecasts demand monthly, the Air Force updates its forecasts only quarterly. Moreover, DLA tracks demand and collaboration at the NIIN level, while the Air Force operates on a weapon system level. Translating between the two levels requires manual steps. Beyond sharing information once demand plans reflect new plans of record, the services may also have preliminary information about prospective changes that are not fully actionable yet but could still be useful for DLA to know. For example, when a need for a product change develops, the owning service typically initiates an Engineering Change Proposal (ECP). This might be done to improve durability, reliability, or capability; reduce cost; or correct a safety issue (which makes it more urgent). ECPs can also be initiated by a contractor suggesting a change, such as by the original equipment manufacturer of the weapon system. The ECP, which contains technical and cost information, goes to the program manager within the service, and a configuration control board makes a decision whether or not to approve it. If approved, the ECP becomes a Design Change Notice, which is used to procure the new NIIN. In order to link the old and new NIINs, an interchangeability and substitutability transaction must then be submitted to DLA Logistics Information Services to update the DoD NIIN catalog. A catalog update enables DLA supply planners and buyers to start replacing the supply of the old NIIN with a supply of the new NIIN. As reported in our interviews, this is an area where communication can be delayed or break down because creating the interchangeability and substitutability transaction is a manual process executed by the services for use by DLA.

The entire process, from when a change is first considered until the design actually changes, can take many months. This is a period in which DLA could conceivably start planning for the item change. However, there is a limit to what DLA can and should do with in-process ECP information because many ECPs are ultimately not approved. There is also a concern that if customers in the field become aware of a possible change in an item, they will begin to look for it before it is ready. Still, interviewees indicated that the time between the Design Change Notice reflecting the approval of the new design and the interchangeability and substitutability transaction reflecting the update of the catalog could be two to three months or longer; this is time in which DLA could safely start the transition to the new item. Additionally, if the DLA supply planner knows that an ECP is awaiting approval or even in the development process, action might be taken to limit order quantities, or at least watch the item that could be replaced more closely, in order to avoid significant long supply if the ECP is eventually approved.

To help compensate for the time involved in executing these many manual steps, DLA applies a manual override on the forecast, using the Digital Demand Dial tool to increase or decrease overall demands. The Digital Demand Dial enables a demand planner to quickly increase or decrease the demand forecast temporarily in response to specific customer or weapon system changes. Once relevant NIINs are identified and the percentage change in demand estimated, the demand planner can increase or decrease the forecast with a matching percentage change for a set period of time. However, this can sometimes be too coarse a tool because, for NIINs with multiple applications, only the portion of the demand associated with some weapon systems may decline and not the other portions.

The need for manual intervention can cause a burden on the people who must convey information to DLA in addition to their primary job responsibilities, as well as on the DLA personnel who must use the information. When operations tempo is high and people are under pressure to complete tasks with more-immediate consequences, this burden can create a further obstacle to information flow. Moreover, because a large number of personnel affect agility, the flow of information depends on a many-link chain, which increases the risk of it slowing or stopping at some point. The wide array of personnel involved can lead to the needed destination of information becoming unclear or even unknown (both reported and seen in examples provided to the research team), although the services and DLA have tried to clarify this with training. Finally, because personnel are dispersed both across and within DLA and the services, their own effect on supply chain agility is not always clear, nor is it tracked.

Recommendations

To address these challenges and increase the flow of demand information from customers to DLA, we recommend further elevating the importance of communicating information about demand changes. This requires senior management–level emphasis in the services, in DLA, and in DoD at large focused on coordination across organizations. Such an enterprisewide emphasis is necessary to account for the wide dispersion of people through whom information flows and to stress that information-sharing is a priority even for personnel who are several steps away from DLA supply planning. While sales and operations planning involves coordination between DLA and each service, effective collaboration between DLA and its customers needs to also occur at the program level.

Part of an enterprisewide emphasis should also involve sharing the cost of inventory excess. Currently, though DLA relies on the services to provide necessary information, DLA bears most of the burden of inventory excess. We recommend an alternative approach, where DLA and the services more directly and transparently share accountability for processes that affect the generation of excess and more transparently share the cost when unneeded purchases are disposed. To implement this, we recommend first that the demand forecast accuracy metric for DLA-managed items also be used to measure service performance, in conjunction with DLA, because customer planning information about demand changes is important for forecast accuracy. This can help balance accountability for processes that affect DLA inventory excess and encourage information-sharing. Where demand forecast accuracy is measured DoD-wide, we recommend separately examining items that are planned collaboratively and other DLA-managed items where one service is the dominant user. Beyond this, highlighting the part of the surcharge that results from inventory excess and making it transparent at joint meetings would also serve to reinforce the need for improved information-sharing.

We also recommend that changes to some current processes be considered. First, we recommend assessing the implications of earlier communication from the services. If the services communicated upcoming NIIN changes earlier, DLA would have the opportunity to slow down purchasing and avoid inventory buildup. However, if taken too far, the services could face a shortage of parts if the replacement NIIN takes longer than expected to get approved or to procure. Assessing these risks would be useful to an accurate weighing of the costs and benefits of earlier information-sharing. Second, where collaboration is currently limited to particular NIINs (e.g., in the Air Force), we recommend assessing workload impacts of expanding this collaboration to all NIINs. This includes workload impacts for both the services and DLA. Third, we recommend evaluating the effects of the Digital Demand Dial on demand forecast accuracy to ensure that it is having the intended effects. We heard some concern in the services that the tool was dialing demand down too far. Fourth, because the flow of demand information requires so many steps and personnel, we recommend exploring possibilities for automation in collaboration efforts. Reducing the number of manual steps and processes could lead to significant increases in accuracy, efficiency, and compliance.

To examine all of these issues in greater detail, we recommend forming an integrated process team with personnel from DLA and the services to first map and then identify improvements to the engineering change information flow process.[2] The team would also identify and examine potential automation opportunities.

Finally, in Figure 3.1, we offer an example of an information exchange that could serve as a repository for upcoming engineering changes across the services. Our discussions with personnel in DLA and the services made clear the need for a process to facilitate the early and consistent flow of information about these changes to those involved in managing each item and to minimize the many steps and people involved. This kind of information exchange application would be a place for Engineering Services to notify DLA and the other services of a possible change to the use of a NIIN at different stages. This would enable immediate sharing with all involved rather than having to proceed through a serial process. When no changes are being considered, the system would code the NIIN green and buying should proceed normally. When a change is possible, the organization considering the change would code the NIIN yellow. This would send an alert to the demand planner, supply planner, and buyer for the item. Then they could alter supply planning and purchasing as determined appropriate. These alterations could include reductions in coverage durations, the elimination of bulk buys, and the possibility of manual reviews of purchases based on discussions among the DLA demand planner and supply planner and the service personnel working on a change. When a change is probable, the engineering organization would code the NIIN red with an alert again being sent out, and DLA would need to check with the program office and Engineering Services before purchasing. At this point, DLA and the service could plan the implementation of the new NIIN. Once a change is complete, the responsible service personnel would code the NIIN gray, and DLA would cease purchasing the old item and make plans to eliminate its inventory. The key maintenance requirement for DLA and the services would be keeping the personnel assigned to the NIIN up to date in the system so that alerts could go to the right person, and so that DLA personnel know whom to contact.

---

[2] This would be in addition to the recent integrated process team addressing first-article testing.

Figure 3.1
Information Exchange for Reporting Imminent Changes to NIINs

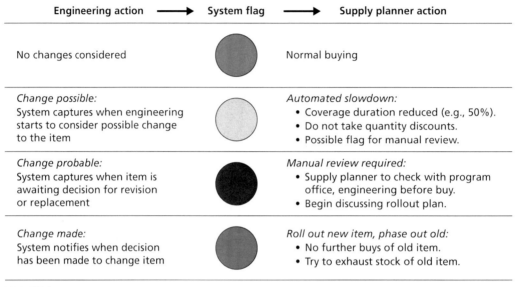

RAND *RR822-3.1*

While customer collaboration has been challenging for DLA, the services have processes for sharing upcoming changes to the demand for a NIIN. However, we also found that there are likely opportunities to improve these processes and to investigate some areas for further improvement. With these changes, we believe that collaboration can improve. Because supply chain agility depends on good customer collaboration, we recommend continuing efforts to improve and expand it.

CHAPTER FOUR

Order Quantities and Agility

An agile supply chain is one that can respond quickly to changes in demand whether they are increases or decreases. Ordering inventory in larger-than-optimal quantities hampers DLA's ability to respond efficiently to decreases in demand. Such order quantities also tie up additional capital and thus can reduce flexibility to respond to changes in demand if a budget constraint has been hit and cannot be adjusted in time.

In this chapter, we discuss opportunities for DLA to more efficiently manage inventory through its choices of order quantities, balancing the costs of investing in inventory (including the risk of excess) against the burden of processing purchase requests, while maintaining materiel availability. However, even optimal but relatively large order quantities are themselves a symptom of a supply chain that is not agile—one that has long lead times and potentially burdensome purchasing processes that cannot efficiently execute frequent procurement actions. Thus, in addition to the direct effect of improving agility by reducing lead times, improving procurement processes can produce a second-order effect of even greater agility and inventory improvement through lower optimal order quantities associated with lower purchase request (PR) costs.

Items that are stocked periodically require replenishment. When the lead times for procuring items are long, orders must be placed well in advance of their need. Forecasting demand over a long lead time is subject to great error even when done as well as possible, with error increasing as lead time increases. Forecasting algorithms can only account for past trends and planned actions that will affect demands, not shifts unaccompanied by customer information provided a lead time ahead. The longer the lead time, the more likely that an event will occur that will change customer demand between the time an order is placed and when its delivery is complete. So, especially with a long lead time for an item, it is possible that demand will decrease between when DLA has awarded a contract for an item and the time at which it receives delivery.

DLA has procedures for identifying when an item that has been ordered is well in excess of its new forecasted demand, which may result in an attempt to cancel the order. However, because a supplier may have purchased raw material or subcomponents and/or begun production, the supplier may have already incurred costs as a result of DLA's order. Consequently, canceling orders often requires a penalty fee that could make it not worthwhile to do so. As a result of having committed to a large order well in advance of the projected demand, DLA could become left with inventory that is never sold and must be disposed of at a loss. Because of large order quantities, DLA is less able to respond to decreases in demand.[1] In contrast, if DLA were able to order materiel in smaller batches, it could more efficiently cope with demand decreases by not having so much on order when demand shifts occur and then further and more quickly reducing the size of orders subsequent to such shifts.

Determining the best order quantity represents a balancing act between two competing objectives. The first objective is to minimize inventory costs, which include the physical storage costs and the opportunity cost of capital, but DLA's case is primarily driven by the risk of inventory obsolescence, and, consequently, the risk of disposal.[2] Meeting this objective would lead to smaller orders placed frequently. The second, competing objective is to minimize the burden on acquisition of processing PRs. Meeting the objective of minimizing PR costs would push for larger orders placed more frequently. To find the optimal balance between achieving the two goals, the workload burden is translated into a monetary cost per PR so that it can be expressed in common terms with inventory costs. This enables determining the order quantity that provides the optimal trade-off between the two costs, providing the lowest total cost.

The commonly employed and well-known approach for computing order quantities is the economic order quantity (EOQ) model.[3] The EOQ formula determines the quantity that produces the optimal trade-off between the cost of inventory and the cost of PRs (see Appendix C for a more complete discussion of EOQ). The balance depends on the relative costs of each, which are parameters that must be estimated for the organization (and that change and should be adjusted as processes improve). If inventory holding costs are high, the model will push order quantities lower, with a resulting increase in the number of PRs. If PR costs are high, the model will do the opposite.

The EOQ model assumes that demand occurs at a constant rate that is known and that orders arrive instantly (i.e., lead times are zero). In real applications, however, lead time is not zero, and the demand in a given time period is not only stochastic but the average often also changes over time. Nonetheless, the model serves as the basis for most order quantity determination methods in industry. Changes in demand rate over time are handled by periodically updating forecasts. Lead times are handled by setting a reorder point so that orders are placed ahead of when they are needed. Stochastic

---

[1]   Similarly, demand for items can increase beyond the forecast when lead times are long, leading to stockouts. DLA has procedures for identifying when existing orders are insufficient to meet the new forecasted demand, and new orders can then be placed to meet the demand. However, depending on the size of the demand increase, the warning time for the demand increase, and the item's lead time, it is possible that the new order will not arrive in time to prevent a stockout.

[2]   A discussion of obsolescence costs and their contribution to inventory holding cost is found in Appendix B.

[3]   See R. H. Wilson, "A Scientific Routine for Stock Control," *Harvard Business Review*, Vol. 13, 1934; and F. W. Harris, "How Many Parts to Make at Once," *Factory: The Magazine of Management*, Vol. 10, 1913.

uncertainty in the demand during that lead time is handled by keeping extra inventory, known as *safety stock.*

However, even these companion methods do not fully account for the differences between the real world in most cases and the EOQ's assumptions. Ideally, the order quantity and safety stock should be computed together, as one affects the other. For simplicity of implementation, however, it is often the case that the order quantity is computed separately from the safety stock using the EOQ. For most industry segments and items, though, this works quite well despite the assumptions that underpin the EOQ, with truly optimal solutions not far from what this produces because of short lead times and other factors.

In the remainder of this chapter, we discuss the DLA method for computing order quantities, which is based on using the EOQ formula but has additional rules, and manual overrides are applied to the computed order quantities. In addition, the implied holding cost assumption has been set inaccurately for reasons we will discuss. We will show that changing the holding cost parameter to one that more accurately reflects DLA's inventory holding cost and removing the extra rules and manual overrides presents opportunities for DLA to reduce both inventory costs and PR workload. This comes through rebalancing inventory and PRs more efficiently among the different items being purchased. We note, however, that efforts to reduce order quantities must also be balanced against the need for inventory to cover demand during the lead time while awaiting replenishments, with a trade-off between order quantities and safety stock requirements to meet service-level goals. This is addressed at the end of the chapter.

How DLA Describes Order Quantities: Coverage Duration

DLA expresses order quantities in terms of coverage duration, or CovDur for short. The coverage duration is the number of days of demand that the order should be able to cover, and thus represents the expected time between orders. Based on the EOQ logic, DLA has constructed a table of coverage durations (which we will refer to as the *CovDur Table,* or *Table*) rounded to 30-day increments and based on the item's forecasted annual demand value (ADV), which is the forecasted annual demand multiplied by its unit price. This closely approximates the EOQ, particularly because the optimal portion of the order quantity curve for items is generally relatively flat (i.e., moving a little bit away from the optimal point has a small effect on costs). At the time the Table was built, there was an emphasis on restricting the inventory investment, which was constrained. To meet the constraint, the computation implicitly assumed a high inventory holding cost that we estimate at 36 percent, which is significantly higher than the 18 percent typically considered DLA's holding cost, resulting in smaller order quantities and a higher number of PRs than would be optimal if 18 percent is close to the actual DLA holding cost.

The coverage duration that is actually found in the inventory management system (the *System*) is based on the Table but modified with added rules and manual overrides. One of the significant rules used is that the coverage duration should be at least as long as administrative lead time; this rule was intended to prevent a second PR for an item from being generated while the first PR is still being processed by a contracting officer. Other rules consist of minimum order quantities and minimum purchase increments, as may be imposed by suppliers. In addition, a manual override capability allows the supply planner to input a different coverage duration than that indicated by the Table or the rules. This might be set in an effort to increase materiel availability or to decrease the PR workload.

Items that are on long-term contracts (LTCs), known at DLA as *outline agreements*, follow a somewhat different rule. LTC items are generally set at 90-day coverage durations, unless the item has a high ADV above $100,000, in which case the coverage duration is set at 30 days. (For convenience, we will refer to this as the 90/30 rule.) Rules and overrides may also be applied to outline agreement items to produce the coverage duration that is found in the System. DLA Aviation uses a modified table for non-LTC items and a different ADV cutoff for the 90/30 rule for LTC items, with the modified table producing increased coverage durations.

Rebalancing Order Quantities Among Items

The rules and overrides that DLA has applied modify the results of EOQ-based calculations and generally have the effect of lengthening the coverage duration, or, equivalently, increasing the order quantity. While there may be good reasons for wanting order quantities to be increased, such as to reduce PRs or to improve materiel availability, over time the use of these added rules and manual overrides has the potential to unbalance the system, putting extra inventory in some items while requiring too much PR workload in others. Meeting the objectives of the manual overrides could be done more efficiently by adjusting EOQ parameters to reflect constraints, such as workload limits, which can be represented as adjustments to cost parameters (e.g., if there is a PR workload constraint, using a PR can be thought of as an opportunity cost that prevents generating a PR for another item). In effect, this approach of adjusting cost parameters until the predicted PR workload becomes feasible would then optimize order quantities across items subject to this constraint.

To examine the potential for rebalancing the inventory and PR workload among items, we compared the coverage durations currently found in the inventory System against the base DLA Table for non-LTC items and the 90/30 rule for LTC items. We also recomputed order quantities employing the EOQ formula, using an annual inventory holding cost of 18 percent of the price of the item, a per-PR cost of $441.55 for non-LTC items, and a per-PR cost of $20.84 for LTC items. These PR cost estimates are the DLA estimates for these types of PRs, and 18 percent is typically considered to be the holding cost percentage. (More details about estimating DLA holding costs can be found in Appendix B. Information about how average inventory and number of PRs change as PR cost and holding cost parameters are adjusted can be found in Appendix C.) We compared the results for the forecastable items (replenishment method code "R"), excluding dual-channel items.[4]

In Figure 4.1, we show the inventory value of the average cycle stock levels, indicated on the left axis, and the number of PRs, indicated on the right axis, for the three order quantity options. *Cycle stock* is the inventory associated with the purchase, and then consumption, of the order quantity. Note that safety stock is not included here; we will come back to this point later. Moving from the System coverage duration values to the Table values—in other words, removing the manual overrides and other rules from the current coverage duration computations—would result in a dramatic decrease in order quantities, and therefore in a decrease in the average inventory investment associated with cycle stock. Unfortunately, given that there is already a PR backlog, doing so would cause an unacceptable increase in the number of PRs. However, moving from the CovDur Table to the EOQ with 18-percent holding cost (lower than

Figure 4.1
Inventory Investment and Purchase Requests for Non-LTC Items

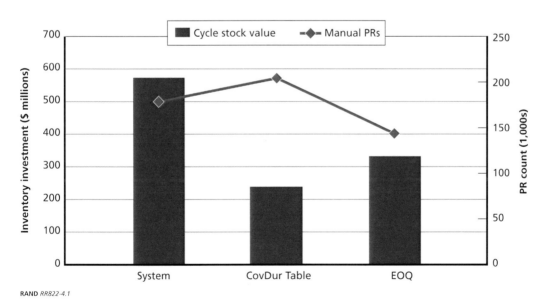

RAND RR822-4.1

4   Dual-channel items are those in which stocking and distribution are split between DLA and the supplier. For some customers, demand is fulfilled from DLA stock, while other customers' orders are fulfilled via direct delivery from the supplier.

that used in the Table) allows for a decrease in PRs at the cost of more inventory than the base Table. The net result, compared with the original System values, is a decrease in both average inventory and the number of PRs.

Ordinarily, inventory and PRs trade off against each other, with a decrease in one causing an increase in the other. The reason that moving from System to EOQ achieves this unusual win-win is because the current System values are unbalanced. They put too much inventory in some items, particularly high-ADV items, while requiring the expenditure of too many PRs on other items, particularly low-ADV items. This indicates that the manual imposition of overrides has not been optimal, increasing both PRs and inventory from the ideal levels. Using the EOQ with an 18-percent (or so) inventory holding cost rebalances inventory investment and PRs among items.

Figure 4.2 shows the ADVs for each NIIN, with the NIINs ranked in order of decreasing ADV. (One NIIN was removed; it was the NIIN with the top ADV, which is so high that it would have distorted the graph, making it look even more like a vertical line and a horizontal line rather than a curve.) The first (relatively) few NIINs have extremely high ADVs, in the millions. Thus, the blue line hugs the y-axis at the left. ADV then tails off rapidly, producing the curved portion of the blue line, with a very large number of NIINs with relatively low ADVs. As a result of the many low-ADV NIINs, the blue line hugs the x-axis. Using the EOQ with 18-percent holding cost and no overrides leads to purchasing the lower-ADV items in higher quantities, accepting

Figure 4.2
Non-LTC NIINs, Ranked by Annual Demand Value

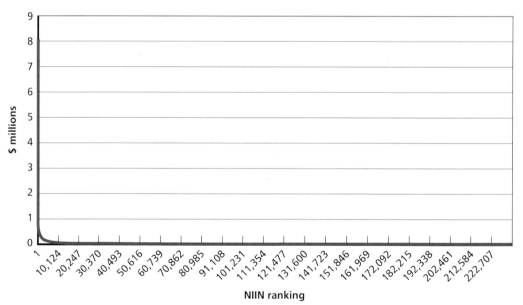

relatively higher inventory levels in inexpensive items (but for small increases in absolute inventory value) in order to free up PR workload that can be deployed to purchase the higher-ADV items in smaller quantities, saving substantial inventory investment. (The DLA CovDur Table would do the same as EOQ, except tuned to a lower overall level of inventory, and consequently a higher number of PRs.)

For LTC items, a different effect is seen. With LTCs, delivery orders are placed using automated systems. This reduces the time and expense associated with purchases, and therefore allows for much smaller and more-frequent orders without affecting workload. Moving from the System coverage durations to the 90/30 rule would reduce average cycle stock, and moving to EOQ would further decrease cycle stock. In each case, the number of PRs would increase, with the average inventory cycle stock and associated PR counts shown in Figure 4.3. However, because the PRs would be executed as automated delivery orders, DLA personnel reported that this would not present a burden, with the $175 million reduction in inventory coming with just a $4 million increase in PR costs, based on DLA's estimate for the cost of an automated order. In the figure, we also present a fourth option, where the EOQ result is constrained to be no less than 30 days of supply; we title this option EOQ30. This option was considered in response to concerns from DLA that purchases executed more frequently than 30 days would be impractical. As can be seen in the figure, the average inventory and the PRs associated with EOQ30 would lie between that of the unconstrained EOQ and the 90/30 rule.

Figure 4.3
Inventory Investment and Purchase Requests for LTC Items, by Policy

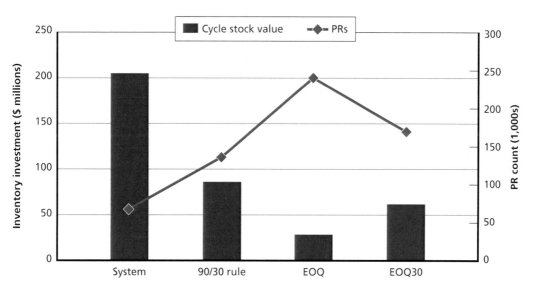

Safety Stock and Order Quantity Trade-Off

Removing the rules and overrides and applying a lower holding cost percentage enables the use of the EOQ formula to reduce both manual PR workload and average investment in the inventory. The inventory affected by order quantities—or in other words, the inventory that results as orders come in and decreases over the course of the purchase cycle as it is shipped to meet customer demand—is known as *cycle stock*. In addition to cycle stock, managers must plan for safety stock, which is maintained to protect against stochastic uncertainty in lead-time demand, whether it be from a higher-than-expected demand rate or from a longer-than-expected lead time. Correspondingly, the factors that affect the amount of safety stock needed to ensure a desired level of materiel availability include the projected demand rate and estimated variability in demand, as well as the length and variability of the lead time.

Another factor that should affect safety stock is the frequency of orders, because backorders will occur toward the end of each cycle, as stock runs low just before replenishments arrive. With smaller order quantities, orders are placed more frequently; consequently, risks of backorders occur more often. In order to maintain the same level of materiel availability, safety stock levels need to increase as order quantities decrease—but not linearly. When considering changes to order quantities, analysts should compute the new safety stock levels corresponding to the new order quantity to ensure that decreases in cycle stock are not fully consumed by the need to increase safety stock. This may happen for high-ADV items that have long lead times and highly variable demand: The high ADV leads to shorter coverage durations, while the long lead time and variable demand lead to high safety stock levels. With these, it may actually be more efficient to improve materiel availability through purchasing such items in larger order quantities. However, this is not true of all items, so it is not a policy to follow blindly. Still, DLA manages many more of this type of item than is common in most private sector industry segments.

Although order quantity and safety stock decisions should be related, they are often computed separately, such as using EOQ to compute the order quantity first and then computing the safety stock that corresponds to that order quantity. Instead, the order quantity and safety stock should be computed together; methods that do so are known in textbooks and in the academic literature broadly as *(Q, R) inventory policies*.[5] More details about these computations and the trade-off between order quantity and safety stock in achieving service-level goals may be found in Appendix D. As noted earlier, for most items and industries, the differences are not significant. Because computing order quantities and safety stock jointly with (Q, R) policies is much more complex, it has become relatively common practice to use the EOQ formula and compute safety stock separately, as is the practice in DLA. However, this difference can be greater for

---

[5]   G. Hadley and T. M. Whitin, *Analysis of Inventory Systems*, Englewood Cliffs, N.J.: Prentice Hall, 1963.

long-lead-time, highly variable demand items more reflective of the DLA population than for most companies and industries.

Order Quantity Recommendations

Opportunities exist to improve both inventory and PR workload by rebalancing order quantities among items using EOQ-based methods. This will require removing the rules and overrides that DLA has applied on top of EOQ-based calculations, as well as employing a holding cost rate that corresponds to an estimate of DLA's actual holding cost. In DLA, 18 percent has been assumed as the approximate holding cost rate, and we estimated 22 or 23 percent in our analysis reported in Appendix B. However, estimating the true cost is fraught with difficulty, with significant uncertainty in our estimate, and excursions we conducted suggest it could be as low as the 18-percent range. The precise rate is not critical; rather, it is important that it be in the right range. So, the rate should be significantly lower than what has been implicitly applied to DLA's order quantity calculations (36 percent). This combination of changes will enable a reduction of cycle stock without increasing PR workload. For LTC items, the use of automated delivery orders enables reduction of inventory costs at low burden even though PRs will increase. For non-LTC items, increases in inventory investment in low-ADV items frees the PR workload needed to buy high-ADV items in lower quantities, resulting in both lower overall inventory costs and lower PR workload on balance. Reducing order quantities on the higher-ADV items allows DLA to more quickly adjust orders downward on these items when demand decreases, reducing the chances of inventory excess when focused on the dollar value of inventory. Reducing PR workload also allows DLA to be more agile by reducing the backlog of PRs and thus reducing administrative lead time. Further improvement may be gained in the future by employing (Q, R) inventory policies.

Further—and Larger—Inventory Improvements Depend on Improving Agility

It is important to note, however, that while the use of EOQ and a (Q, R) inventory policy would offer some improvement, larger reductions in inventory investment depend on improving the processes associated with supply chain agility. The reason that some of the overrides were implemented, increasing order quantities, was an inability to keep up with the PR workload. While rebalancing workload among items helps reduce the workload, a more effective solution would be to streamline the purchasing process, making more-frequent, smaller order quantities more efficient and effective. The use of LTCs and automated orders helps, but setting up these contracts is a long and labor-intensive process, and this can be applied only under certain conditions—relatively high and predictable demand.

Some of the inventory gains achieved by reducing order quantities are counteracted by necessary increases in safety stock. To some extent, this is a necessary and unavoidable trade-off that is inherent in the relationship between order quantities and safety stock. However, the extent to which safety stock increases may outweigh decreases in cycle stock depends on the amount of safety stock needed. Here, too, the agility of the supply chain comes into play in determining this trade-off. Safety stock depends on the uncertainty in the demand over the lead time. While improvements in forecasting and in information-sharing with customers can reduce that uncertainty, they will never eliminate it, particularly in the defense context, where demand for many items will change over time and where demand for many NIINs has a very high level of variability even when stable.

Instead, reducing lead time, both administrative and production, presents the greater opportunity for improving forecasting and thus reducing inventory levels, because there is less error in forecasting demand when the forecast horizon is shorter. With a shorter lead time, lead-time demand variability will come down, reducing safety stock, and the chance of a trend shift leading to excess inventory or a customer support shortfall will go down. As a result, with shorter lead times, as order quantities are brought down, the need for increased safety stock will also go down, increasing the potential for and the benefit of reducing order quantities.

The next chapter, on acquisition processes, focuses on reducing lead times, among other improvements, while also considering factors to ensure that lead-time reduction does not come at the expense of negative trade-offs in other processes and costs, as discussed in Chapter Two, so that total cost and performance improve.

CHAPTER FIVE

Acquisition Processes

The acquisition community is a key player in supply chain agility. In this chapter, we describe findings related to acquisition processes and the resulting recommendations aimed at elevating the emphasis on supply chain agility and better integrating it into acquisition processes. These include reducing PLT; considering supplier performance, PLT, and order quantities along with purchase price in bid selection; expanding LTCs; and continuing to reduce ALT. These findings are based on detailed interviews with personnel in multiple acquisition roles at DLA's three supply chains (Aviation, Land and Maritime, and Troop Support) and at DLA Headquarters, supplemented by a review of DoD and DLA policy on purchasing and supplier management, particularly as it relates to lead-time management. Detail about our interview methods can be found in Appendix A, and the policy review is covered in Appendix F. The recommendations also draw on industry best practices derived from a literature review and interviews with selected firms, as described in Appendix E.

Working with Suppliers to Reduce Production Lead Time

Reducing lead times is a central component of supply chain agility, one that is often of concern in the private sector. To this end, negotiating with suppliers for reductions or continuous improvement in PLT is a best practice in the commercial sector that appears applicable in a government context (see Appendix E for a discussion of this practice and its use by leading companies). However, we found that DLA's acquisition processes are, by and large, not focused on reducing PLT.

Supervisors appear to put little emphasis on PLT. For example, they do not regularly ask buyers to use PLT as part of bid selection. Buyers appear to perceive that PLT is largely dictated by suppliers for contracts, and they therefore perceive that they have little negotiating power, especially for the many items that are unique to the military because there is often a limited supply base and market for these items. Even for commercial items, there is a lack of PLT benchmarking information, which contributes to buyers' perception that suppliers largely dictate PLT. Thus, while they recognized that shorter PLTs were better and chose them when all else was equal, buyers reported having little choice but to accept the PLTs offered by suppliers, as long as they did not differ widely from the historical PLT of record. This is likely, at least in part, because buyers lack PLT benchmarks and information about suppliers' abilities to reduce PLT. For most items, the PLT used for comparative purposes in contract negotiation is the historical PLT of record, which is the PLT of the previous contract for the item. We found similar results with respect to order quantities, with buyers perceiving that suppliers dictate quantities and that they have little room for negotiating. Further, it is not clear that the full value of shorter lead times and lower order quantities is broadly understood, and, as will be discussed later, buyers lack the tools for making this value assessment and comparison. Finally, DoD and DLA policies do not emphasize reduction in PLT as a goal (see Appendix F).

Rather than focus on comparing or trying to reduce PLT, DLA acquisition personnel focus on delinquencies and compliance with contracted delivery times.[1] This is important in supply planning, and over time, it is valuable to reduce the PLT as well. Emphasizing only on-time delivery can hinder supply chain agility because it creates an incentive for suppliers to contract for lengthy PLTs to ensure that they will meet the on-time delivery requirements. In contrast, leading commercial firms have found success in reducing PLT by focusing on it as a continuous improvement goal in conjunction with measuring on-time delivery.

Given the significance of PLT for both inventory risk (the length of PLT) and supply planning (compliance with planned or contracted PLT), we recommend that DLA add PLT as a continuous improvement goal for suppliers, alongside the current on-time delivery goal. This combination would encourage suppliers to contract for and meet shorter lead times, with contracts having shorter PLTs over time, which could either be within an LTC or from contract to contract. The natural place to begin using PLT as a continuous improvement goal is with suppliers with which DLA has established a *strategic supplier alliance* (SSA). SSAs are arrangements that DLA puts in place with some vendors with which it has more than $100 million in sole-source business. DLA and these suppliers have a high degree of day-to-day interaction to facilitate customer support. Acquisition staff can also work with other suppliers, especially those with which DLA does substantial business, such as during the renewal of an LTC.

We note that driving for shorter PLT should not come at the expense of higher prices that outweigh the value to be gained by the shorter PLTs. Hence, we call it a continuous improvement goal to imply a focus on process improvement rather than paying higher prices for lower PLTs. This is also an example of why, in Chapter Two, we emphasize the need to employ the full set of outcome metrics that often involve trade-offs. Later in this chapter, we will discuss the need for a tool to help buyers weigh

---

[1]   Even with the compliance focus, tracking delivery times and issuing sanctions or penalties when late to ensure compliance during postaward contract management are reported as limited.

the value of differences in PLT versus differences in price when a lower PLT bid comes with a higher price.

In addition to serving as a continuous improvement goal, adding PLT as a metric will also generate more PLT data in order to develop benchmarks. Especially where items are not commercial, objective benchmarks are difficult if not impossible to find. This is exacerbated by infrequent ordering for the many parts that have low demand. Our discussions with acquisition personnel indicate that PLT goals are based largely on past PLT, especially for military-unique parts. Relying solely on past PLT performance presents potentially long PLTs as benchmarks. This is especially the case for items with low demand. Thus, PLT should also be tracked over time to see if there has been improvement, with goals set for continuous improvement over time as contracts are renewed or rebid.

Broadening the Definition of *Best Value*

Firms aim to reduce PLT because it produces value, either in improved inventory efficiency or improved customer service. Considering PLT becomes part of determining which supplier and bid will be the best value. But related to the limited focus on PLT, we found that DLA acquisition personnel, both those interpreting policy and those writing and executing contracts, tend to think of and apply the best value clause of the Federal Acquisition Regulation (FAR) primarily in terms of price. Acquisition personnel reported that they include supplier performance as part of best value, but to a limited degree. This is because easily useable information about supplier performance is insufficient, and acquisition personnel do not have the tools for converting the available or prospective information about supplier performance to how it affects value. In particular, many expressed dissatisfaction with the data available on supplier past performance. Measures of supplier performance are limited to those in the federal Past Performance Information Retrieval System (PPIRS)—and these are primarily subjective, with ratings like *very good*, *satisfactory*, *marginal*, and *unsatisfactory*—to cover areas such as cost control, management responsiveness, and product performance. The measures do not include PLT.

In the long term, we recommend that DLA increase the available supplier performance data to include PLT performance over time and standardize its use in supplier selection for contracts with higher demand value (i.e., use should be limited to contracts where the additional workload is more likely to be worth the effort through value gained by DLA). First, DLA should determine the desired measures and then work with the Defense Procurement and Acquisition Policy (DPAP) to get these requirements integrated into PPIRS. At a minimum, we recommend adding PLT and updating delinquency data as soon as the due date passes. With the needed data defined, the process developed, and data collected, DLA can then implement a more comprehensive, systematic approach to including supplier performance in supplier selection from a total value standpoint.

Acquisition staff members do not incorporate PLT or order quantities—and the associated disposal risks—into the definition of *best value*. This is partly due to a limited understanding of how long PLTs contribute to inventory risk and cost and to a lack of tools to account for the risk of disposal. As a result, we recommend explicit DLA policy that defines the FAR's best value clause to include PLT and order quantities, in addition to price, and that provides acquisition personnel with the tools to account for the value of these parameters. Including PLT and order quantity in the supplier selection criteria and in the contract negotiation can reduce the risk of stockouts and excess inventory. It is also a best practice for reducing lead times in the commercial sector. Where items are commercial or dual-use, commercial PLTs for the items could be used in the negotiation. To accomplish this, acquisition staff would need to identify sources of such commercial PLTs, either individually or in databases. Negotiating PLT and order quantity in conjunction with price may require additional training for buyers, as part of a larger program on supply chain agility and its effects on cost and performance.

Thus, we also recommend developing and standardizing the use of a tool to help buyers and supply planners balance the inventory risk associated with long PLTs and large order quantities with an item's purchase price to determine the lowest total cost. The recommendations in the prior paragraph cannot be implemented without such a tool. Typically, supply-planning policies and systems set order quantities, but buyers and supply planners are periodically presented with the possibility of paying a reduced price if a larger quantity is ordered. In these cases, buyers propose higher order quantities to supply planners based on the buyers' interactions with suppliers, and supply planners either confirm or reject the usability of the higher quantity based on demand expectations. However, in their current calculation of whether the bulk buy would be the best strategy, supply planners and buyers lack the means to incorporate the risk of excess inventory and eventual disposal if demand unexpectedly drops and falls short of projections.

As a result, we recommend designing a tool to systematically account for trade-offs among prices, order quantities, and lead times. Such a tool would estimate the total annual cost of various alternatives, based on lead times, order quantities, prices, annual demand, the cost of completing the PR, and the holding cost rate. Figure 5.1 shows an example of such a tool. Annual purchase costs are computed based on the annual demand and the unit price, including whatever quantity discounts are offered by suppliers. Annual PR costs are computed based on the annual demand, the order quantity, and the estimated DLA cost per order; the order quantity could differ from the EOQ depending on the requirements for receiving the quantity discount. Annual holding cost for cycle stock is affected by the differences in unit price and order quantity. Annual holding cost for *pipeline stock* (stock necessary to cover the expected demand

Figure 5.1
Example Tool to Weigh Alternative Total Annual Costs

Baseline				Offer A	Offer B
Fixed cost of PR	440				
annual demand	500				
holding cost rate	18%				
unit price	$ 100.00	Alternative price/unit offered:		$ 95.00	$ 102.00
EOQ	157	Order quantity required:		250	157
lead time (wks)	26	New lead time (wks)		52	15
Annual cost with EOQ:		Annual cost with qty discount:			
annual purchase cost	$ 50,000.00	annual purchase cost		$ 47,500.00	$ 51,000.00
annual PR cost	$ 1,401.27	annual PR cost		$ 880.00	$ 1,401.27
annual holding cost, cycle stock	$ 1,413.00	annual holding cost, cycle stock		$ 2,137.50	$ 1,441.26
annual holding cost, pipeline	$ 4,500.00	annual holding cost, pipeline		$ 8,550.00	$ 2,648.08
TOTAL annual cost	$ 57,314.27	TOTAL annual cost		$ 59,067.50	$ 56,490.61

RAND RR822-5.1

during the lead time) is affected by the differences in unit price and lead time, and it is included to reflect the risk of a demand change that would lead to this material being on-hand and potentially excess.

In the example in Figure 5.1, we compare the base option using the EOQ against two options. Offer A has a lower price but a higher required order quantity and longer lead time; Offer B has a higher price but a shorter lead time. In this case, Option B is shown to be the best strategy because it is projected to have the lowest total annual cost, after weighing lead times, order quantities, and price.[2]

Most of the inventory value risk comes from high-ADV items—that is where the money is.[3] Once a tool is prototyped, implementation that incorporates the valuation of lead times and order quantities for supplier and bid selection on a standard basis should begin with items that have the highest ADV, starting with a relatively high threshold and then lowering it to a point to be determined. We note that DLA Land and Maritime has developed a similar kind of tool focused on order quantities and quantity discounts, which could be adapted to also incorporate lead times. Currently, using the Land and Maritime tool is optional.

---

[2] The prototype tool shown here could be further extended to include the effect of order quantity and lead times on safety stock. Increased order quantities have the potential to reduce the amount of safety stock required, because with fewer orders per year, there would be fewer times per year in which inventories run low. Thus, increased cycle stock could be mitigated somewhat by the decrease in safety stock. In contrast, longer lead times, often associated with increased order quantities, increase the uncertainty in the demand during the lead time, which increases the amount of safety stock necessary. Computing these precisely would require more data on the demand for the item, as well as the desired service level. More information on how order quantities affect safety stock is found in Appendix D.

[3] Recall that ADV is the unit price times the annual demand for an item.

Expanding Long-Term Contracts

Lead times and order quantities can also be reduced—and supply chain agility increased—by placing items on LTCs. LTCs are associated with lower lead times and enable smaller orders in several ways. Once the contract is written, ALT drops to almost zero because orders placed against the contract are automated; the ALT is taken off the critical path. LTCs are also associated with shorter PLTs, though it is not clear whether this is a function of the LTC or of the items placed on LTCs. In addition, when items are placed on LTCs, acquisition workload is freed to work on PRs for items that are not on LTCs (though this comes with more upfront workload to establish the LTCs). Order quantities naturally become lower because the cost per order is much, much lower. Additionally, smaller order quantities are also associated with shorter lead times. Due to these benefits, separating the sourcing decision (i.e., contract development) from the buying process (i.e., placing orders against the contract) that is enabled by LTCs is a commercial best practice for reducing lead times (see Appendix E). See Appendix G for a full discussion of the relationships among lead times, contract types, item categories, and other factors.

DLA has been working to increase the prevalence of LTCs for some time. Purchases of items on LTC accounted for 7–18 percent of the NIINs and 22–43 percent of the dollars spent between September 2010 and August 2012, depending on the supply chain. (See Figures G.23 and G.24 in Appendix G.) Interviews with DLA personnel suggest that DLA, during the course of this project, has further increased the emphasis on placing items on LTCs, and we recommend continuing these efforts with additional, strategically planned growth. In particular, our analysis found that placing NIINs with the highest ADV onto LTCs produces the most value. Doing so would relieve more PR workload, compared with items with lower ADVs, because they are ordered most frequently. It would also provide the greatest reduction in inventory investment from the smaller, more frequent orders. Our discussions with acquisition personnel indicate that concentrating on high-ADV items is generally DLA's practice, but explicitly using ADV provides a convenient and consistent template for identifying LTC candidates, and our analysis confirms the value of DLA's practice (see Appendix H for the analysis). Other factors—including demand variability and price volatility— affect the feasibility, supplier interest, and value of using LTCs; these can be considered once the ADV-based candidate list is developed.

Continuing Efforts to Reduce Administrative Lead Time

Finally, our discussions with acquisition personnel pointed to several ongoing efforts to reduce ALT, as well as to the need for continuing and expanding these efforts. The Time to Award team, which began after we briefed initial findings to DLA Headquarters, significantly accelerated these efforts. The DLA director stood up the team in April 2013 to identify ways to reduce the time between a PR and the contract award. This extensive effort has produced many recommendations, which reinforce the findings and recommendations here, such as the need for more pricing tools, more-streamlined regulations, and more-detailed process improvements. Fully implementing the team's recommendations promises to produce meaningful improvement in ALTs.

Further emphasized by the Time to Award team, an ongoing effort to reduce ALT has been to streamline duplicative acquisition regulations. We heard about these efforts in several places. Continuing this work can reduce ALT by enabling buyers to work more efficiently, especially where those regulations increase touch time or calendar time. DLA buyers work under four layers of acquisition regulation (federal, DoD, DLA, and DLA supply chain), and sometimes these regulations overlap. Ideally, these multiple layers add clarification for purchasing in particular contexts. More often, buyers reported that the multiple layers add more time than clarification. Many buyers described a recent increase in the amount of regulations, and especially an increase in DLA and DLA supply chain regulations, that duplicate existing policy and add both touch time and calendar time to contract development. For example, one supply chain regulation requires a screen shot to confirm and create an audit trail proving that the winning bidder is not disbarred, and additional legal and signature reviews are required at some supply chains (for lower contract levels than required by DoD or DLA Headquarters policy). While any one regulation might add only a few minutes of work time to developing a contract, when multiplied over the course of the many contracts written at DLA, the result can be significant.

We also learned that DLA is working more closely with the services to develop and gain compliance with Memoranda of Understanding regarding Engineering Services turnaround times, which can reportedly add months to ALT. Reducing these turnaround times could be fruitful for reducing lead times. In addition to this ongoing effort, we recommend that DLA and the services increase the emphasis on supply chain agility within the Engineering Services community, particularly by increasing the visibility of and management emphasis on their turnaround times, which should be shown and discussed in joint forums.

Another effort that reduces ALT is DLA's expanded use of automated purchasing, which is a process where solicitation, bidding, and awarding are automated, requiring no manual involvement of a buyer and substantially reducing ALT. The automated purchasing application has price thresholds to identify bids that are inordinately high, defined as a certain percentage higher than the previous purchase price (e.g., 25 percent). If all bids have prices above the price threshold, then it drops out of the automated process, and the contract is completed manually, which drives up the ALT. Price thresholds thus need to strike a balance between minimizing lead times with the possible acceptance of excessive price markups from a threshold that is too high and creating more manual purchases that cost more in workload to execute than the value to be gained from possible reductions in prices from a threshold that is too low. We recommend regularly evaluating price increase thresholds used for automated purchasing to maintain this balance.

Finally, DLA has additional ongoing efforts to reduce ALT, including eliminating the backlog of PRs, combining more items on contracts, expanding the use of online contract approvals when contract officers are working away from the office, and redistributing the acquisition workload (e.g., by grade or function) to increase efficiency. Many of these are addressed in greater detail by the Time to Award team's recommendations.

Conclusion

Our discussions with DLA acquisition staff and our analyses of lead time data indicate several areas where DLA could likely increase supply chain agility by incorporating lead times and order quantities more fully into acquisition processes, particularly by adding increased emphasis on reducing PLT. Emphasis on reducing ALT has already dramatically increased over the last year and a half.

First, DLA should work with suppliers to reduce lead times by adding PLT as a continuous improvement goal along with the current on-time delivery goal. This can also serve to develop a database of PLTs for military-unique items to develop better PLT benchmarks to establish these goals. Second, DLA should integrate lead times, order quantities, and supplier performance into supplier and bid selection by clarifying the definition of *best value* from the FAR and how it should be applied in DLA. In conjunction, DLA should expand the use of supplier performance data in determining best value, although DLA will first need to work with DPAP to gain the inclusion of the needed data in PPIRS. DLA will also need to further develop and clarify practices on how to integrate supplier performance data into the value determination process. Additionally, as part of applying this broader notion of best value, DLA should develop and employ a tool to help supply planners and buyers best balance increased inventory risk from higher lead times and order quantities with lower material prices when such trade-offs come into play. Third, DLA should continue efforts to strategically expand the use of LTCs and reduce ALT. To build on these recommendations, the next chapter provides the suppliers' perspective on how DLA could work with them to improve agility.

CHAPTER SIX

Supplier Perspectives

We interviewed DLA suppliers to identify additional opportunities for improvements in DLA's processes to reduce costs or lead times, as well as to compare our other findings with suppliers' perspectives. We interviewed ten DLA suppliers, all of which have a large volume of business, a large number of contracts, and at least some LTCs. The suppliers spanned all three DLA supply chains (Aviation, Land and Maritime, and Troop Support) and included five SSA suppliers and five that are not SSA suppliers. Two of the ten suppliers are certified small businesses. Complete details about the interview methodology can be found in Appendix A.

An important but not surprising finding that resulted from the interviews is the range of supply chain maturity among the suppliers. Differences that revealed this range include how suppliers use information about demand, how they work with customers, and how they manage their own suppliers. As a result, their responses about how DLA could work with them to improve overall supply chain agility, including what information and other practices may be useful to them, likely reflect their degree of supply chain maturity. The information and practices that DLA could follow that would help its suppliers improve efficiency or performance in support of DLA depend on the various suppliers' capabilities and processes. For example, the level of sophistication of a supplier's processes may affect the type of information that it can use to plan inventory and production—or has even thought about using—and thus what level of information-sharing it desires. In addition to supply chain maturity, several other factors emerged that relate to how DLA should pursue supply chain agility with its suppliers, including pursuing paths tailored to the needs of different types or categories of suppliers, which we describe below.

Dual-Use Items

The potential for improved supply chain agility appears to be much higher for dual-use items than for military-unique items for several reasons. *Dual-use items* are defined here as those that have both commercial and military use and for which suppliers have both commercial and military customers. Production volume tends to be substantially higher and more stable for these items because of the broader customer base and volume that typically accompany commercial industry. As a result, suppliers report being willing to stock dual-use items, which leads to shorter lead times for the customer (in this case, DLA), greater potential for rushing high-priority orders (i.e., suppliers redirecting resources when high priorities emerge), and greater interest in potentially managing inventory for DLA.

Because of these multiple benefits for supply chain agility, we recommend maximizing the use of dual-use items across DoD where possible when it provides similar or close-to-similar capability. Because the services decide what items can fill a requirement, this recommendation necessitates the services' participation. It includes requiring the services to consider commercial items to fill new requirements and to review practices and regulations to identify any that might hinder incorporating dual-use items. The recent DLA First Destination Transportation and Packaging Initiative may facilitate the use of dual-use items.[1] Because, as our interviews confirmed, an item can become military-unique just from marking or packaging requirements (which may result from regulations or simply from traditional practice), this initiative may lead to more dual-use items by allowing commercial packaging.

Long-Term Contracts and Pacing of Orders

LTCs facilitate supply chain agility, as discussed previously. Most suppliers that we interviewed reported wanting more LTCs as long as they are accompanied by guaranteed minimums high enough to make the workload required to set up an LTC worth it. Most also reported that they would welcome longer terms for LTCs. The large setup costs and long calendar time involved in initiating an LTC discourages suppliers from pursuing those that are too limited in scope. We therefore recommend increasing LTCs where appropriate, with incentives of guaranteed minimums (but not so high as to create undue risk) and longer durations where possible. This is consistent with recommendations made earlier and lends further support to DLA's ongoing effort to increase LTCs.

We heard one exception to increasing LTCs. Suppliers operating in industries with high price volatility do not welcome LTCs, because pricing is too hard to estimate with any accuracy, creating excessive risk to them. Because suppliers in these industries prefer one-time contracts, all items in these industries may not be good candidates for LTCs. We also asked about employing LTCs for low-demand items. Suppliers reported that combining low-demand items with high-demand items in an LTC would be more attractive than developing an LTC for low-demand items only. This was true even if

---

[1] See Defense Logistics Agency, "First Destination Transportation and Packaging Initiative (FDTPI)," web page, undated.

enough low-demand items were grouped together to enable a sufficient guaranteed minimum. Therefore, we recommend grouping low-demand items into LTCs with high-demand items where feasible (e.g., by identifying items produced in the same industry or by working with small-business advocates).

Related to using LTCs, suppliers varied in their preference for having smaller, frequent orders or larger, infrequent orders. Many suppliers pointed to the needs and constraints of their own supply base as determining which ordering mechanism would be preferential. They reported being more concerned with having a guaranteed minimum quantity over the life of the LTC rather than with having one particular sizing and pacing of orders. Thus, we recommend that DLA contracting staff work with suppliers to make the best use of their production and management systems to minimize costs and lead times. This requires understanding suppliers' capabilities and the factors that make them efficient and then customizing order quantity considerations to them. Because these factors can vary by item (e.g., because of differences among subtier suppliers), this recommendation requires the participation of DLA buyers on a contract-by-contract basis, as well as the participation of supplier relationship managers for SSA suppliers.

Customer and Supplier Relationships

Most suppliers we interviewed had at least some commercial customers in addition to DoD customers. Compared with commercial customers, they consistently reported that DLA orders require more administrative work but yield less volume and less profit. Suppliers reported providing shorter lead times to commercial customers in these situations, given the higher yield that the suppliers receive from them. Thus, for suppliers with commercial customers, DLA may offer a smaller and less profitable relationship than commercial customers offer. Given this, we recommend that DLA combine contracts where possible to reduce transaction costs to become more competitive with its suppliers' other customers, enabling DLA to demand better service.

The DLA suppliers we interviewed reported that they negotiate regularly with their own suppliers for shorter lead times. The methods they employ range in sophistication and success. For example, many DLA suppliers reported that they found their own suppliers to be capacity constrained when they try to work with them to reduce lead times, especially for items that are made to order (e.g., military-unique items), which limits improvement. In contrast, one DLA supplier described working with its suppliers to identify where in an item's supply chain the capacity constraints were tightest and then working together to reduce those constraints. For example, if one subcomponent of an item or one type of production material has a particularly long lead time, stocking that subcomponent or production material could prove cost-effective if it enabled the overall lead time for the item under contract to be reduced at minimal cost and risk. Most DLA suppliers did not describe this degree of sophistication in working with their own suppliers. Another variation in working with subtier suppliers stemmed from the level of demand for an item. DLA suppliers expressed reluctance to press their own suppliers for lead-time reductions on items with very low demand for which finding a supplier had been difficult.

Given the overall tendency of these DLA suppliers to work with their own suppliers on lead times and their range of supply chain maturity in doing so, as concluded from industry best practices, we reiterate our earlier recommendation that DLA also use lead times as a key factor in the bid and supplier selection process. For SSA suppliers, we reiterate our recommendation that supplier relationship managers work interactively with those suppliers to reduce lead times. Focusing on SSA suppliers, at least in the beginning, enables the highest return with limited DLA acquisition resources because these relationships are already established.

Other Supplier Factors Affecting Supply Chain Agility

We also asked suppliers about the possibility of redirecting orders when priorities change. Suppliers reported doing this wherever possible but that they sometimes face capacity limits that preclude this, especially when the order swapping is not accompanied by any additional funds. Capacity constraints that limit suppliers' abilities to redirect an order include equipment constraints in manufacturing firms or in subtier suppliers, labor constraints in small businesses, and business volume at the time of the request (i.e., when business is slow, swapping orders is more possible). Suppliers reported that requests for expediting orders from DLA are not uncommon but are rarely accompanied by additional funding and are virtually never accompanied by a *FAR code DX*, which requires the order to be placed at the front of a supplier's queue for military necessity.

Requests for expediting generally come from DLA customers through DLA customer account specialists, who then contact DLA supply planners and contracting officers. DLA contracting officers could facilitate these requests for changes in priority by working together and with supply and demand planners to identify how multiple orders placed with the same supplier for items managed by different contracting officers should be prioritized to balance customer needs and then communicate this to the supplier. However, suppliers reported that this does not happen often. We thus recommend that DLA encourage contracting officers and supplier relationship managers to coordinate with each other and the relevant supply and demand planners in order to prioritize orders. Whoever first becomes aware of a need to expedite an order should check to see what other items the supplier is providing to DLA, and DoD more broadly, and then initiate the coordination.

In our interviews, we also asked about information-sharing with suppliers because commercial literature suggests that real-time demand and inventory information-sharing should facilitate supply chain agility. Indeed, this is a common best practice in commercial purchasing.[2] The suppliers we interviewed generally concurred, although there was variation in their responses that seemed to mirror their level of supply chain sophistication. For example, some suppliers simply recognized that better demand information would be useful but is difficult for DLA to obtain, while others asked for specific kinds of information, and still others sought access to information directly from the services. In general, demand information that is better and more frequently updated would reportedly be useful for suppliers. This includes historical information on use and failure rates and better intelligence on the services' needs. Two suppliers that had worked directly with the services reported better demand information from the services than they receive from DLA, which underscores the need for DLA and the services to improve customer collaboration. As this collaboration improves, we recommend increasing information-sharing with suppliers, particularly of the improved forecasts that result. This would improve communication and coordination, two mechanisms for improving collaboration and supply chain agility.[3] In addition, we recommend that DLA provide portals that enable suppliers to access data as desired, which will help account for the range in supplier needs and capabilities. These data should include real-time inventory, historical demand, and forecasts.

We also asked suppliers about the potential for managing more of DLA's inventory—that is, increasing the amount of direct vendor delivery business with suppliers (i.e., customer-direct). Increasing this type of business for dual-use items appears to be very attractive to suppliers because it is a natural extension of their practices with commercial customers and builds on that inventory. In contrast, increasing the number of military-unique items on direct vendor delivery appears to have limited appeal to suppliers because of the substantial increase in inventory risk that they would face. This shifting of risk would also involve increased purchase prices.

Taken together, it appears that several aspects of supply chain agility can be improved by working more closely with DLA suppliers based on their needs and capabilities. DLA can reduce lead times by

- collaborating with the services to pursue the use of more dual-use items
- using more LTCs where appropriate
- using lead times in the bid and supplier selection process
- working with suppliers to reduce their lead times
- identifying the most efficient sizing and pacing of orders for each supplier

---

[2] See, for example, Daniel R. Krause and Robert B. Handfield, *Developing a World-Class Supply Base*, Tempe, Ariz.: Center for Advanced Purchasing Studies, 1999.

[3] Gligor and Holcomb, 2012a.

- coordinating orders within DLA among items sourced from a supplier when an order needs to be filled quickly
- improving forecasting information and data availability for suppliers.

Order quantities and inventory risk can be reduced by many of these processes as well, including encouraging direct vendor delivery with more dual-use items, expanding the use and scope of LTCs, and identifying the most efficient sizing and pacing of orders.

CHAPTER SEVEN

Conclusions and Recommendations

Supply chain agility is the responsiveness and efficiency with which customers are served when they have changing needs. Improvements in supply chain agility enable an organization to reduce materiel stockouts and inventory excess that leads to disposal. Improving three processes enables supply chain agility to improve: increasing the speed and accuracy of the delivery of information about planned changes that will affect customer demands, shortening lead times, and reducing order quantities.

DLA has strong incentives for improving its supply chain agility. It places a high priority on avoiding stockouts to meet customer needs, but to achieve this, it has been generating excess inventory, resulting in an average of $1 billion in annual disposals. Improving supply chain agility offers opportunities to increase customer service and inventory efficiency, thereby reducing the annual level of disposals and thus the annual level of purchases for inventory.

Through analyses that spanned the entire DLA supply chain, this project investigated how DLA might improve its supply chain agility. Based on the results of these analyses, the project identified a number of recommendations for process improvements that will lead to greater supply chain agility. DLA is already implementing some of these recommendations.

Increase the Emphasis on DLA Supply Chain Agility Enterprisewide

The central finding and recommendation of this project is a need for increased enterprisewide emphasis on supply chain agility. The three main supply chain outcomes—materiel availability, inventory turns, and purchase price—are determined by interacting processes that are affected by multiple functions and organizations across the supply chain. This means that performance metrics, whether for outcomes or processes, must be applied from an end-to-end supply chain perspective. The only place where all outcomes come together in DLA is at the DLA director level. Because the services also have an influence on forecasting and administrative lead time, the only place these outcomes and factors all come together is at the DoD supply chain enterprise level. Thus, an increased emphasis on supply chain agility requires involvement from the most-senior management levels across the DoD supply chain management enterprise and flowing downward to all levels.

Our first recommendation, therefore, is for strong management involvement across the enterprise at all levels. Such involvement has three elements: ensuring the integration of supply chain agility into all processes, adding new performance metrics and enhancing some existing ones, and training all stakeholders. Process integration involves not only policy and procedures but also day-to-day engagement—for example, asking agility-related questions in briefings and reviews to emphasize agility and continually force all personnel to consider how they can affect it. New metrics include

- using inventory turns for all items in the DLA monthly APR
- using award price change compared with the producer price index in the APR
- adding ALT, PLT, and order quantity overrides to the Inventory Management Council
- measuring the services, in conjunction with DLA, on demand forecast accuracy for DLA-managed items.

Additionally, monitoring disposals over time as a percentage of sales and should-be inventory would provide an overall view of whether agility is improving (in conjunction with continuing to measure materiel availability and ensuring it stays the same or improves). Fully training all communities on supply chain agility ensures that they understand its importance and how they affect it, enabling them to combine the motivation to improve agility with the knowledge to do so.

Continue and Expand Efforts to Reduce Lead Times

Our second recommendation is to continue to shorten lead times. DLA's efforts to reduce ALT have expanded since the start of this project, and we recommend continuing these efforts and implementing the DLA Time to Award team's recommendations. The next step is to expand such efforts to PLT reduction (as an additional charter of the DLA Time to Award team). We recommend that DLA work with its suppliers to reduce PLT by tracking it and using it as a continuous improvement metric, with performance improvement favorably affecting supplier selection. DLA should also work with its major suppliers to identify what sizing and pacing of orders would enable the lowest costs and shortest lead times. In addition, we recommend that DLA combine contracts where possible to lower transaction costs for suppliers and coordinate internally how to prioritize multiple orders with the same supplier if one order becomes a high priority, and then convey that information to the supplier.

We recommend that DLA use PLT to a greater degree and in a more standard way in bid selection. We found that buyers have some understanding of the importance of PLT but lack the guidance, data, tools, and thorough understanding of supply chain agility to fully address it in their processes. Regarding guidance and policy, this means explicitly defining *best value* to include lead times and order quantities, in addition to price. Regarding data, we recommend that DLA work with DPAP to expand the data in PPIRS, the data system on supplier performance, to include PLT. To effectively implement *best value*–focused policy and employ improved data, buyers and supply planners need a tool that weighs the value of competing options by accounting for trade-offs among prices, order quantities, and lead times, particularly when faced with a price break for a bulk purchase. Figure 5.1 in Chapter Five offers an example of such a tool. Finally, we recommend maximizing the use of dual-use items, which have shorter lead times, where possible. This would require a new form of collaboration with the services to encourage dual-use item selection when practical, because they make the decisions on which items to use.

Right-Size Order Quantities

Our third recommendation is to right-size order quantities by finding the optimal balance of purchasing workload and inventory levels to minimize total cost. To minimize the risk of purchases that lead to inventory disposal, ideally, order quantities are small, especially for items with long lead times. However, because processing more small orders is also associated with acquisition labor, the two needs must be balanced. The current DLA system for determining order quantities, or the coverage duration of an item, has become unbalanced by using manual overrides, and the implied inventory holding cost is putting too much weight on minimizing inventory—thus leading to the DLA supply chain overrides to reduce the high workload burden. But manual overrides do this in a suboptimal way. We recommend using the EOQ with an inventory holding cost that closely approximates the estimated DLA holding cost, adjusting the cost-per-PR parameter as necessary to apply PR workload constraints. This rebalances both inventory levels and the number of PRs (i.e., the acquisition labor involved) by decreasing the size of order quantities for high-ADV items and the number of PRs overall (primarily through PR reductions for the very large number of low-ADV items). In the longer term, safety stock levels should be computed jointly with order quantities in order to fully optimize inventory.

Continue Efforts to Increase the Use of Long-Term Contracts

Our fourth recommendation is for DLA to continue to expand its use of LTCs, especially with guaranteed minimums, longer lengths, and where low-demand and high-demand items can be combined in order to incentivize suppliers to increase the use of

LTCs. Most suppliers appear to want more LTCs, especially when accompanied by guaranteed minimum orders and longer contract lengths. Items in industries with high price volatility are an exception to this recommendation because suppliers in those firms may be reluctant to commit to a long-term price. Delivery orders for items under LTCs are associated with lower lead times because the supplier will have already been selected, reducing the cost of placing each order, which in turn facilitates smaller order quantities. Thus, prioritizing NIINs with the highest ADV for placement on LTCs realizes the greatest benefits.

Improve the Flow of Information About Demand Changes from Customers

Our fifth recommendation is to improve the flow of information about upcoming changes to NIINs from customers to DLA. Supply chain agility increases when information about upcoming changes in demand is transmitted more quickly and accurately. DLA and the services have processes in place to collaborate on demand planning, especially with the depots. Another area in which tighter collaboration would be valuable is when Engineering Services changes the usage of a NIIN such that a retired NIIN needs to be phased out and replaced by a new NIIN. Effectively conveying this information enables DLA to avoid oversupply of the old NIIN and undersupply of the replacement NIIN. However, the number of personnel and steps involved in conveying information about engineering changes are high and the process is complicated, which slows the flow of information or makes it unclear who should receive such information.

This again points to the need for management-level emphasis and process redesign, both within and across the organizations, and for training about how individual roles affect supply chain agility. We recommend creating an integrated process team to examine engineering change and information-sharing processes for further improvements. This could include examining how process changes and collaboration between DLA and the services could enhance dual-use item selection to also help shorten lead times.

Finally, developing a repository for exchanging information about engineering changes may be one way to give DLA information at different stages. It could designate the likelihood of change for a NIIN from no change being considered to potential change in development to change likely or in process. DLA could then use this information to improve decisionmaking and reduce risk. For example, identifying NIINs that might be changed would enable DLA to scrutinize bulk buys more carefully to avoid purchasing inventory that could become excess. As collaboration between DLA and the services improves, we recommend sharing this information with suppliers by providing portals through which suppliers can access forecasts, inventory levels, and other information as their needs and capabilities dictate.

Potential Directions for Further Improvements in Supply Chain Agility

These recommendations, which span the supply chain from top to bottom and from end to end, can facilitate DLA's supply chain agility. As we discuss below, though, these recommendations will not fully exploit the potential of enhanced supply chain agility but rather represent a strong start to improving it, emphasizing the most urgent areas of supply chain agility for DLA to begin or to continue addressing.

Early in this report, we noted that supply chain agility involves being able not only to better respond to changes in demand but also to better handle shifts in supply. However, given prior research and the most visible problems pointing to the critical need to address the ability to handle demand shocks and significant shifts in trends at the item level, this was the focus of our project and this report. The improvements to address demand changes naturally involve substantial focus on supply responsiveness. Some of these efforts should not only improve responsiveness but also help DLA begin improving awareness of supplier issues that could result in delays or disruptions—or even new sourcing opportunities. This should come through improved supplier management, tighter supplier integration, and increased scrutiny of supplier capabilities as part of determining best value. The next step in improving supply chain agility, though, would be to more deliberately target increasing awareness of looming supply issues and opportunities and to be able to effectively handle supply problems. This could be done through future projects, whether conducted internally, such as by a Tiger team, or jump-started with external help.

Additionally, much of the focus of this report was on item-level demand changes instead of on understanding and reacting to the broader environment, where changes could have implications for major segments of items, items related to a particular weapon system, or even much of the item population that DLA manages. Alternatively, major changes in the environment could have implications for major portions of the supplier base or the supply of certain types of items, such as those based on a raw material that faces a global capacity crunch. With regard to the demand environment, based on the authors' broader interactions with DLA, there appear to be relatively good tie-ins to the rest of DoD to ensure that DLA is aware of major operations being considered that could have significant implications for the support it must provide. So too are there mechanisms in place for adjusting forecasts across broad segments of items in such cases. Still, systematic review of the processes—with regard to being alert to and being able to effectively integrate DoD planning information from program to combatant command levels—could bear fruit in further improving DLA support effectiveness. On the supply environment side, there is likely opportunity for improvement from better intelligence on supply segments as a whole, particularly given limited DLA visibility and integration beyond first-tier suppliers.

So, the recommendations provided in this report reflect merely a first step on the journey toward enhanced supply chain agility. The journey should continue beyond implementing these recommendations for continuous improvement by tackling supply chain agility more comprehensively, which promises to produce substantial financial and customer service gains.

Interview Methods

We interviewed 147 people about DLA processes related to supply chain agility. The people we interviewed served in a wide range of capacities across the supply chain. Table A.1 shows their general roles and the number of personnel interviewed in each group.

Interviews followed a common approach. Before the interview, we prepared questions. During the interview, one or two people took detailed notes while another person led the questioning. After the interview, we compiled and analyzed notes and vetted findings.

Table A.1
Characteristics of Interviews, by Community

DLA Community	Number Interviewed
Supply chains	
Acquisition	20
Supply planners	30
Demand planners	29
Business processing community	8
Other (customer-facing personnel)	2
National account managers	8
DORRA staff	5
Other headquarters staff in J3 and J7	28
Customers working in collaboration	2
Suppliers	15 personnel at 10 firms
Total interviewed	147

NOTE: DORRA = DLA Office of Operations Research and Resource Analysis; J3 = DLA Logistics Operations; J7 = DLA Acquisition.

Selection

Sampling strategies for the interviews took several forms. Supply chain personnel were selected in two ways. Most interviewees were selected by senior DLA personnel based on our requests. We requested and completed interviews with acquisition personnel who had experience with one-time purchases, LTCs, or SSAs. In the planning communities, we simply requested and completed interviews with demand and supply planners. We also requested interviews with examples of other personnel involved in aspects of supply chain agility, such as customer information flow and order quantity determination. From this request, we spoke with business processing and customer-facing personnel. In all functional areas, we spoke with staff in a range of levels, from operational to supervisory to division chief, including persons in GS-9 through GS-15 positions. All of these interviews were conducted in person.

The other way that we selected supply chain personnel to interview was based on particular NIINs that we identified as fitting certain types of demand and supply patterns. The project includes analyses of sample NIINs with histories pointing to problems or successes in supply chain agility. Interviews with the demand and supply planners assigned to these NIINs helped us map the processes involved in reacting to abrupt changes in demand. These interviews were conducted via telephone.

Most other interviews with DLA staff involved questions about particular roles or processes in the supply chain. Personnel for these interviews were selected based on their functional roles. We conducted a group interview with eight national account managers or their deputies to learn about their role in customer information flow. Similarly, we spoke with headquarters staff involved with customer account specialists and other customer-facing personnel about customer information flow. We spoke with relevant staff in DLA Materiel Policy, Process, and Assessment (J33) about coverage durations, economic order quantities, and other logistical operations. We spoke with relevant staff in DLA Acquisition (J7) about acquisition operations and with DLA Office of Operations Research and Resource Analysis (DORRA) staff about their workload study and about coverage durations. We spoke with headquarters personnel involved with performance metrics and with those involved in two specialty teams, the Tiger team and the Time to Award team. Some of these interviews were in person, and some were via telephone.

We interviewed DLA customers and suppliers. We interviewed personnel in two of the three services who work with DLA customer collaboration. These persons were selected based on their functional roles, and the interviews were conducted on the phone.

The DLA suppliers we interviewed came from ten companies. We interviewed 15 people in ten interviews. The suppliers were selected primarily based on volume of business with DLA. We identified companies that were among the top 100 in terms of dollars and contracts with DLA in each of the three supply chains. We then asked the supply chains to review the list to suggest suppliers to interview. Two supply chains offered alternate companies to some on our suggested list. At Troop Support, some companies primarily engaged in direct vendor delivery, which would not give us needed insight into supply chain agility. Troop Support then provided us with high-volume suppliers of stocked items. Land and Maritime provided alternatives to some of the suppliers we suggested without explanation. The supply chains provided contact information for the suppliers we interviewed. We solicited interviews from 17 companies, and ten responded to our interview requests. We sent multiple follow-ups to the remaining seven companies, but they were unwilling to complete the interview. Five of the ten companies interviewed were SSA suppliers; five were not. Two of the firms were certified small businesses; eight were not.

Finally, we interviewed industry leaders in supply chain agility as well. Details about those interviews can be found in Appendix D.

Interview Methodology

As noted, most interviews with supply chain personnel and some interviews with head-quarters personnel were conducted in person. Where this was not possible, interviews were conducted via telephone. In the supply chains, 57 of the people interviewed were interviewed in person; 32 were interviewed via telephone. Interviews with headquarters personnel are less easily classified into in person or telephone because many people participated in more than one interview, some of which were in person and some on the phone. Interviews ranged from 30 to 90 minutes, depending on the length of the protocol and the number of people in the interview. There were one to six people in each interview.

During each interview, one to two people were designated notetakers who wrote as detailed notes as possible, compiling near-transcripts of each interview. This method avoids the risk of missing important information by saving all editing and analyzing until later. Typical note-taking, which focuses on summarizing the information that is most important, runs the risk of overlooking details whose importance only becomes clear later, when interviews are analyzed together or other results come to light. This approach also avoids interviewee hesitation and reduced responses that can occur with recording, especially with interviews that are relatively short and based on job responsibilities.[1]

To maximize the information we received, we promised interviewees that we would not quote them directly nor include their names or titles in written reports. The sampling strategies precluded us from promising anonymity. However, we do not believe that this limited the information we received, because we asked open-ended

---

[1]   R. S. Weiss, *Learning from Strangers: The Art and Method of Qualitative Interview Studies*, New York: Simon and Schuster, 1995; I. Seidman, *Interviewing as Qualitative Research: A Guide for Researchers in Education*, New York: Teachers College Press, 2012.

questions to learn about practices and how processes are executed, not about controversial topics.

The protocols of interview questions were developed based on relevant DoD and DLA policy, relevant outside literature, and results of the study to date. Protocols for acquisition personnel, DLA suppliers, and commercial industry leaders drew on academic literature of commercial best practices. Protocols for the supply planning and demand planning communities drew on logistics theory and practice, and protocols for all personnel drew on results gleaned to date. Protocols are included at the end of this appendix.

Analysis

Once a group of interviews was complete, individual researchers read the transcripts of each group of interviews (e.g., demand planners, suppliers) and identified themes in that group. We looked across all responses to the same question and within each interview across questions. Where applicable, we also grouped interviews by supply chain to identify possible supply chain differences. We vetted themes in three stages with different members of the research team to reduce the risk of bias from any one interpretation. This approach mirrors other qualitative research.[2]

For interviews where there was not a whole group of interviews, such as the national account managers or the Time to Award team, we went through a similar process. We identified themes individually, vetted them as a research team, and then compared the themes with other results in process.

We also vetted all findings with DLA to clarify where findings might need further investigation or a different emphasis and where our resulting recommendations were already being implemented. In some cases, we followed up with additional interviews to clarify or add further information to a theme.

The resulting findings and recommendations reflect interviews across multiple groups and settings. As with any qualitative analysis, care should be taken not to overgeneralize and presume that every theme is universal. This is especially the case with the supplier interviews and the customer interviews, where the number of interviews was small and there was less opportunity to vet findings with related groups. At the same time, we have minimized the risk of biased results by interviewing multiple groups in multiple locations, by analyzing near-transcripts of the interviews, and by vetting the results through a research team and with DLA.

---

[2]   See, for example, Seidman, 2012; Weiss, 1995; N. W. Jankowski and K. B. Jensen, *A Handbook of Qualitative Methodologies for Mass Communication Research*, New York: Routledge, 1991; and A. K. Daniels, "Self-Deception and Self-Discovery in Fieldwork," *Qualitative Sociology*, Vol. 6, No. 3, 1983.

Protocols

Interview Protocols for DLA Staff for RAND Study, "Supply Chain Agility" Sponsored by the Assistant Secretary of Defense for Logistics and Materiel Readiness

We have been asked by the Assistant Secretary of Defense for Logistics and Materiel Readiness in coordination with the Defense Logistics Agency (DLA) J3 and J7 to identify opportunities for improvements to DLA's purchasing and supply chain management practices that are focused on improving lead times, reducing order quantities, and mitigating the effects of shifts in demand on inventory planning in order to lower total costs. The purpose of this interview is to gain the information we would need to fully understand DLA purchasing and inventory management processes. We will not reveal the names of persons we interview. Rather, we will aggregate our findings across all of our interviews and characterize our findings in a general way.

Can you walk us through the following processes?
We are especially interested in how these processes are related to lead times, order quantities, and staff members' workloads and in any related policies or guidance that you routinely use.

Demand Planners

1. Please describe your job
   a. Job title
   b. How long in this position
2. SLIDES with example NIINs
3. Process for demand planning
   a. Process for integrating customer demand data
   b. Process for getting information about expected changes in customer demand
   c. Process for identifying or verifying unexpected changes in customer demand
   d. Process for adjusting automated forecasts
   e. Process for balancing DLA's financial priorities with customers' priorities
   f. Process for working with other logistics personnel: supply planners, customer account specialists
   g. How does EBS [the Electronic Business System] help or hinder these processes? What parameters are adjustable?
4. Process for determining the right mix of people (e.g., supply, demand, acquisition)
5. Suggestions of customers to interview
6. Phased delivery plans
7. Performance measures on which someone in your position is evaluated

Supply Planners

1. Please describe your job
    a. Job title
    b. How long in this position
2. SLIDES
3. Process for supply planning
    a. Process for dealing with exceptions
    b. Process for assessing or prioritizing items
    c. Process for changing the supply plan when there is a change in the forecast
    d. Process for dealing with large, unanticipated, unusual orders from customers
    e. Process for balancing DLA's financial priorities with customers' priorities
    f. Process for manually adjusting what system recommends, if possible
    g. Process for working with other logistics personnel: demand planners, customer account specialists
4. Process for determining the right mix of people (e.g., supply, demand, acquisition)
5. Suggestions of customers to interview
6. Phased delivery plans
7. Performance measures on which someone in your position is evaluated

Acquisition

1. Please describe your job
    a. Job title
    b. How long in this position
2. SLIDES
3. Process for completing a PR
    a. When the item is already on an outline agreement (long-term contract)
    b. When an item is not on an outline agreement (long-term contract)
        ◦ Micro purchases
        ◦ Other purchases
    c. Supplier selection process
        ◦ Market research process, if any
    d. Process to prequalify suppliers, if any
    e. If the item is part of the Strategic Materiel Program
        ◦ Policies and guidance, plans for expanding
        ◦ Data and information shared
        ◦ Is the Strategic Materiel program evaluated internally (for individual items or overall)? If yes, please explain. How do you know it is improving acquisition?
    f. Process for incorporating small businesses or for meeting small business goals
    g. Process for determining types of contract used

    h. Process for determining incentives and penalties on contracts

    i. Process for setting order quantities on contracts
- Are these set by the supplier?
- Process for requesting exceptions to DoD's 2-year order quantity

    j. Process for determining delivery quantity vs. order quantity

    k. Inclusion of lead times in contracts

4. Process for evaluating supplier performance (metrics, which suppliers to prioritize)

    a. How is supplier performance used?

5. Process for obtaining technical review

6. Process for establishing a long-term contract

    a. How an item is identified for a long-term contract or assigned to one

    b. How that contract is negotiated

7. Processes for reducing administrative lead time or production lead time

8. Process of working with strategic supplier alliance companies

    a. Policies and guidance, plans for expanding

    b. Do these processes address PLT, ALT, risks of delays, supplier problem solving, supplier improvement? Please explain.

    c. Data and information shared

    d. Is the SSA program evaluated internally (for individual companies or overall)? If yes, please explain. That is, how do you know it is improving acquisition?

9. Any other improvement initiatives in acquisition

10. Suggestions of suppliers for us to interview (do not need to be in SSA)

11. Process for determining the right mix of people (e.g., supply, demand, acquisition)

12. Performance measures on which someone in your position is evaluated

**Interview Protocols for National Account Managers
for DLA Suppliers for RAND Study, "Supply Chain Agility"
Sponsored by the Assistant Secretary of Defense
for Logistics and Materiel Readiness**

We have been asked by the Assistant Secretary of Defense for Logistics and Materiel Readiness in coordination with the Defense Logistics Agency (DLA) J3 and J7 to identify opportunities for improvements to DLA's purchasing and supply chain management practices. The purpose of this interview is to gain the information we would need to fully understand the constraints facing DLA's customers and what, if any, process improvements could be made to improve total supply chain cost and performance. To gather this information, we are interviewing personnel throughout DLA's supply chain, including customers, suppliers, and DLA personnel at headquarters and at the Aviation, Land and Maritime, and Troop Support supply chains.

1. Please describe your role.
2. We're interested in learning how information flows from the [service responsibility] to demand planners. Is this a process that you facilitate?
    a. If yes, please describe.
    b. If no, can you suggest who we should talk to about this?
3. How is information about major events/program changes that could affect demand provided to the DLA?
4. How do you coordinate with [service responsibility]?
5. Please describe the process by which you learn about upcoming changes in demand.
    a. What weapon system, parts, or groups of parts or organizations/types of activities do you learn about?
    b. What weapon system, parts, or groups of parts or organizations/types of activities do you NOT learn about?
    c. What information do you learn?
    d. What organization gives you the information?
    e. When do you learn the information; how close to the change in what's needed?
    f. Under what circumstances do you communicate this information to DLA? Are there things that would make it easier to provide this information?
    g. Are there policies that describe this process?
    h. Are there automated systems that are currently used to communicate information? Please describe.
        ◦ Do they help?
        ◦ What would make them more helpful?

6. Please describe the collaborative planning process.
    a. What works well in this process?
    b. What could work better? How?
    c. Are there policies that describe this process?
    d. Are there metrics for tracking how well this process works?
7. Are there things that you do or that you encourage people to do to help get parts in quickly?
8. We're trying to put numbers to all the different elements of administrative lead time. Do you know if there are data on the time that it takes Engineering Services to complete their work? If not, can you suggest someone who might know?

Interview Protocols for Order Management, J331
for DLA Suppliers for RAND Study, "Supply Chain Agility"
Sponsored by the Assistant Secretary of Defense
for Logistics and Materiel Readiness

We have been asked by the Assistant Secretary of Defense for Logistics and Materiel Readiness in coordination with the Defense Logistics Agency (DLA) J3 and J7 to identify opportunities for improvements to DLA's purchasing and supply chain management practices. The purpose of this interview is to gain the information we would need to fully understand the constraints facing DLA's supply chain personnel and what, if any, process improvements could be made to improve total supply chain cost and performance. To gather this information, we are interviewing personnel throughout DLA's supply chain, including customers, suppliers, and DLA personnel at headquarters and at the Aviation, Land and Maritime, and Troop Support supply chains.

1. Please describe your role.
   a. Which customer-facing roles do you work with?
   b. Which customer-facing roles do you not work with?
      ◦ Who should we talk to about these?
2. Please describe the role of the customer-facing personnel you work with.
3. We're interested in learning how information flows from DLA's customers to the demand planners. Is this a process facilitated by the customer-facing personnel you work with? If yes, please describe the process by which you learn about upcoming changes in demand.
   a. How is information about major events/program changes that could affect demand provided to DLA?
   b. How is information about one-time or short-term changes in demand provided to DLA?
   c. What information do you learn?
   d. What organization gives you the information?
   e. When do you learn the information; how close to the change in what's needed?
   f. Under what circumstances do you communicate this information to demand planners? Are there things that would make it easier to provide this information?
   g. Are there policies that describe this process?
   h. Are there automated systems that are currently used to communicate information? Please describe.
      ◦ Do they help?
      ◦ What would make them more helpful?

4. Please describe the collaborative planning process.
    a. How does the process work? What does it cover?
    b. In addition to collaborating on demand forecasts, do customer-facing personnel collaborate with customers on how customers order? This could include working with customers on order quantities, the phasing of delivery, schedules for new items, use of substitute parts.
    c. What works well in this process?
    d. What could work better? How?
    e. Are there policies that describe this process?
    f. Are there metrics for tracking how well this process works?

**Interview Protocols for DLA Customers
for RAND Study, "Supply Chain Agility"
Sponsored by the Assistant Secretary of Defense
for Logistics and Materiel Readiness**

This is a study for OSD (Logistics and Materiel Readiness), looking at the Defense Logistics Agency. One aspect of the study is seeing whether there are ways in which DLA can better receive and make use of information coming from the services. One example of that is when there are engineering changes, safety-of-use messages, or other types of changes that may cause one NIIN to replace another. These changes can result in DLA ending up with an oversupply of the old NIIN and an undersupply of the new NIIN. We would like to get some perspectives from people from the services, to learn about the process by which information is conveyed and hear where the process works or doesn't work.

    We will not reveal the names of individual customers in written reports. Rather, we will aggregate our findings across all of our interviews and will characterize our findings in a general way.

1. What is your role? How long have you been in this role?
2. We'd like to learn about the process for making changes to NIINs. But before we start, we want to be more careful about our terminology when we talk about changes.
   a. When a change is made to an item, does that result in a new NIIN being assigned?
   b. So strictly speaking, it is not "changing the NIIN" but "changing to a new NIIN"?
3. Can you tell us more about the [service] organizations that make changes to NIINs?
   a. My understanding is you're at [division within service], which handles sustainment. But there are also the program executive officers and project managers, who are on the acquisition side and report to [commander authority]. Do they make changes to NIINs?
4. Can you walk us through the process of how decisions are made on engineering changes?
   a. Under what circumstances are they made? For instance, here are three changes we know of (there may be more):
      ◦ upgrade to improve the *capabilities* of a weapon system
      ◦ upgrade to improve *reliability* of a weapon system
      ◦ safety-of-use

    b. How are those changes conveyed to those in charge of the items' inventories?

       ◦ How is it conveyed within the [service] to item managers?

       ◦ How is it conveyed to DLA? Which organization is in charge of communicating with DLA?

          – Are there automated systems in place for communicating?

          – Is there a formal process of discussing with DLA the retirement of the old NIIN and the rollout of the new NIIN?

          – How much warning would DLA get? How could DLA be given advance warning, even if a decision on a potential change was not yet final?

5. Are there actions that DLA could take that would help the [service] implement changes?

6. Do you get pushback from DLA on changes in the usage of NIINs?

    a. For example, DLA may have a lot of stock in the old NIIN that it would like to exhaust, before switching over to use of the new NIIN.

    b. How do these differences get resolved?

**Interview Protocols for DLA Suppliers for RAND Study,
"Supply Chain Agility"
Sponsored by the Assistant Secretary of Defense
for Logistics and Materiel Readiness**

We have been asked by the Assistant Secretary of Defense for Logistics and Materiel Readiness in coordination with the Defense Logistics Agency (DLA) J3 and J7 to identify opportunities for improvements to DLA's purchasing and supply chain management practices. The purpose of this interview is to gain the information we would need to fully understand the constraints facing DLA's suppliers and what DLA could do differently to improve total supply chain cost and performance in collaboration with its suppliers. To gather this information, we are interviewing approximately 20 DLA suppliers, chosen based on the amount of business they do with DLA. We will not reveal the names of individual suppliers in written reports. Rather, we will aggregate our findings across all of our interviews and will characterize our findings in a general way.

1. How long have you worked with [company name]? What is your job title/responsibilities? For how many of these years have you worked with DLA as one of [company's] customers?
   a. Do you manage contracts with DLA? If so, how many DLA contracts do you manage?
   b. What are some of the major types of items that you manage?
2. What sizes and pacing of orders work best for you? For example, how do smaller and more-frequent orders compare with larger and less frequent orders?
3. What type of information from DLA would best help you plan your production and orders from your suppliers?
4. Are there opportunities to swap priorities within your production planning so that you could fulfill a DLA order more quickly if it was truly necessary? Are there practices that DLA could provide to help make this kind of priority swap happen?
5. Do you work with your suppliers to shorten the time it takes them to fill an order (that is, their lead times)? If yes, please describe those practices.
6. Do you have long-term contracts (LTCs) with DLA? If yes, consider the following: Some suppliers manage a customer's inventory for them, usually targeted to a certain materiel availability. Other suppliers would rather have inventory management remain the customer's responsibility.
   a. Rather than simply filling orders received from DLA, what advantages (if any) would you find from filling customer-direct orders?
   b. Rather than simply filling orders received from DLA, what advantages (if any) would you find in managing DLA's inventory (i.e., being responsible for keeping it within a specified range)?

    c. What disadvantages would there be to managing DLA's inventory rather than simply filling orders?

7. What else could DLA do that would enable you to shorten your lead times, efficiently take orders and deliver in smaller quantities, or become more efficient in other ways? For example, other ordering behaviors or types of contracts.

8. Do you have commercial customers?

    a. If yes, are your lead times for commercial customers generally shorter, longer, or about the same as the lead times for DLA?

    b. If shorter or longer, please explain why.

APPENDIX B

Estimating DLA Holding Costs

Inventory holding cost is a construct used to represent the costs of keeping inventory. It is used in analytic models to trade off the costs of keeping inventory against the benefits of doing so. Estimating inventory holding costs is important to two sets of decisions: when to purchase inventory and when to dispose of already-purchased but apparently excess or obsolete inventory. As we describe in this appendix, though, the estimate of the holding cost used for each type of decision should not be the same.

A guiding principle is that only those costs that can be affected by the decision should be included in the decisionmaking. Costs that cannot be affected by the decision, or *sunk costs*, should not be included. Following this principle, the holding cost that should be applied when making decisions on future purchases should be different from the holding cost that should be applied when making decisions on disposals. Decisions on future purchases concern the potential commitment of money toward purchasing inventory that may lose its value. In contrast, with disposal decisions, money has already been committed; therefore, the costs associated with this possible loss are sunk. The only remaining cost for an item in this state is the continuing cost associated with holding the item in the warehouse or distribution center.

In this appendix, we describe the types of costs that compose holding costs, discuss which ones are relevant for each type of decision, and compute a rough numerical estimate of the resulting holding cost that should be used by DLA. We also explain why the estimates given can only be rough and why that may be sufficient for decision-making and planning.

Holding Costs for Purchase Decisions

The decision to purchase and hold inventory of an item enables DLA to effectively support its customers, with more inventory improving this service. Purchasing a larger amount of inventory per order, as opposed to a smaller amount, also reduces DLA's purchasing workload. In addition, suppliers may give a quantity discount for larger orders, particularly in cases where there are significant manufacturing setup costs, and thus economies of scale associated with larger production runs.

However, committing to holding inventory comes at a cost. One cost is that money is tied up in inventory and therefore unavailable for other purposes. For a business, this is modeled as an opportunity cost of capital that is based on the internal rate of return that it would expect to earn on other investments; in other terms, the opportunity cost of capital is the weighted average cost of capital consisting of the weighted average of the cost of the business's debt (i.e., current interest rate) and the expected equity returns of its shareholders. The Office of Management and Budget provides guidance on setting the opportunity cost or discount rate to use for decisionmaking for various investments. It employs the principle of tying the choice of rate to treasury rates of maturities that correspond to the investment horizon. It also publishes the nominal Treasury interest rates for different maturities, updating this annually for use by government agencies.[1] The cost of capital for inventories—based on nominal interest rates for three-year Treasury notes and bonds, which would be the term that most closely corresponds to the turnover of inventory—is quite low: 1 percent for 2014.

A much larger cost of holding inventory for DLA is that substantial amounts of the items held in inventory will never be sold, due to declines in demand that were unforeseen when the items were purchased. DLA is then left with inventory that it can never hope to sell in any reasonable amount of time (or ever), and, as a result, disposes of this obsolete inventory at a loss. As discussed in Chapter One, this is a substantial cost to DLA.

The other potentially substantial cost of holding inventory is the storage cost. Storage cost for an item depends on its size and any special conditions under which the item must be stored. Note that we apply the marginal cost to store inventory, which is the operating cost to run facilities and operate host installations. In many situations, the cost of the land and the building itself would be part of this marginal cost, either as a lease or as an asset with an annual depreciation expense. For DoD, the cost of the land and built facilities is sunk, although one could consider the value of selling them as an opportunity cost to be amortized, along with the need to periodically replace buildings. We do not include this cost here, but it could be estimated, although doing so would only be relevant to the extent that an installation could be shed, making the tie to any specific purchase indirect.

Typical algorithms that utilize holding cost to determine an order quantity apply it as a percentage to the unit price of the item. Thus, while the likelihood of obsolescence depends heavily on the type of item being considered, and while storage costs depend more on size of the item than on cost, in order to maintain compatibility with these algorithms, we will likewise estimate holding cost as percentage of the unit price. This is standard academic and industry practice. In this section, we will show that the major DLA costs of holding inventory are value lost to obsolescence

---

[1] Office of Management and Budget, *Guidelines and Discount Rates for Benefit-Cost Analysis of Federal Programs*, Circular A-94, October 29, 1992 (Appendix C, revised December 26, 2013).

resulting in disposal and the storage cost. We therefore compute the holding cost rate for purchases as:

$$\text{Nominal interest rate} + \frac{\text{Disposal value} + \text{Storage cost}}{\text{Inventory value}}.$$

Other costs also used by industry include insurance and shrinkage. The storage costs could include warehouse operations (labor and material), security, and utilities, among any other costs associated with facility operations.

Disposal Value

The amount of inventory that has been disposed by DLA has varied over the past several years. Figure B.1 shows the inventory value of DLA-managed items in serviceable condition that were sent to DLA disposition services.

The large variability in disposals from year to year presents a problem for computing holding costs because the estimate for inventory planning would change wildly from one year to the next if used to change holding costs every year based on the previous year's costs. Additionally, disposals do not typically come from material purchased in the prior year but rather from two, three, four, five, or even more years before. This makes using only the prior year's disposal value an unreliable estimate. An estimate

Figure B.1
DLA Serviceable Item Disposals, 2005–2013

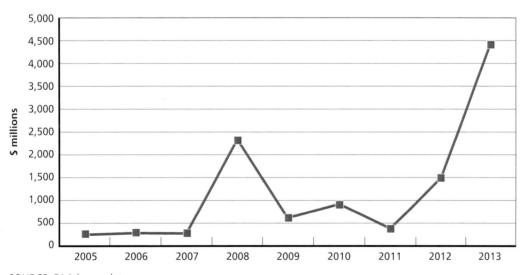

SOURCE: DLA issues data.

RAND *RR822-B.1*

derived from multiple years of data would lead to a more stable and accurate estimate of annual disposal value associated with any year's purchases, and, therefore, holding costs. For the purposes of this example, we used a four-year moving average due to the availability of data that enabled us to break down the disposal data by DLA supply chain, but a different length or method for computing an estimate could be used if desired. The results, by supply chain, using a four-year moving average for 2010–2013 are shown in Table B.1.[2]

Notice that the inventory value of disposals varies by supply chain. This is the result of many factors, including the original value of the materiel, differences in obsolescence rates among different types of goods, and differing levels of forecasting error for different types of materiel depending on such as factors as lead times and demand variability.

Storage Costs

Based on the costs reported in the DLA Working Capital Fund Budget documents, we estimate that total storage costs are $350 million per year for both DLA-owned and service-owned materiel. For the purposes of estimating DLA holding costs, we wish to focus only on DLA-owned and managed items, so these total storage costs need to be adjusted downward to remove service materiel. While the budget documents do not break down the costs by service versus DLA materiel, data are available on cubic

Table B.1
DLA Disposal by Supply Chain, 2010–2013

Supply Chain	Average Value of Disposed Items ($ millions)
Aviation	309.9
Construction and Equipment	27.9
Clothing and Textiles	22.4
Industrial Hardware	20.4
Land	58.8
Medical	3.5
Maritime	115.7
Unidentified	92.6
Total	651.2

---

[2] As stated in the report, the four major DLA supply chains are Aviation, Land and Maritime, Troop Support, and Energy. However, we have analyzed the supply chains by subcategory as well. For example, Industrial Hardware and Construction and Equipment are subparts of Troop Support.

space used by each type of materiel. DLA-owned materiel percentage of cubic space has increased from 45 percent to over 50 percent between 2011 and 2013.[3] From this information, we decided to apply a factor of 50 percent against total storage costs to estimate the cost of storing DLA-owned and -managed materiel. Therefore, we estimate that storage costs for DLA-owned and -managed items are about $175 million per year.

For the purposes of computing holding costs by supply chain, we allocated these storage costs among the supply chains according to an estimate of the space used. We did this by dividing the average on-hand inventory cube by supply chain by the total on-hand inventory in terms of cubic feet of material, using numbers from fiscal year 2011. This produces a supply chain percentage of cubic storage space consumed. Then these percentages are multiplied by the storage cost to develop individual supply chain storage costs as summarized in Table B.2.

Inventory Value

The holding cost we are estimating is one to use in an inventory model that determines how much inventory should be held. DLA then buys according to this plan, but ultimately accumulates more inventory than needed or planned because of demand declines or unrealized plans that are used in forecasts. The disposals that this excess leads to are excess to the modeled level—the holding cost is used in the model, which produces an associated plan, or *should-be* inventory level. Thus, we use the value of the should-be inventory as the denominator in calculating holding cost percentages that are to be used to inform future purchases of inventory (Table B.3).

Table B.2
DLA Storage Costs by Supply Chain, Fiscal Year 2011

Supply Chain	% of Cube	Cost of Storage, by Supply Chain ($ millions)
Aviation	23	40.7
Construction and Equipment	16	28.0
Clothing and Textiles	19	32.8
Industrial Hardware	2	3.8
Land	30	51.8
Medical	1	2.4
Maritime	9	15.5

---

[3] The analysis of inventory stored in DLA Distribution Centers uses the Quantity by Owner file.

Table B.3
Should-Be Inventory Estimate from Fiscal Year 2011 Opening Position

Source	Value ($)
Safety level from requirements column	1,249,210,895
50% of procurement/EOQ from requirements column[a]	1,235,595,064
Insurance objective from requirements column[b]	1,080,398,307
War reserve from on-hand column	96,338,635
Diminishing manufacturing sources from requirements column	140,125,404
Total	3,801,668,305

SOURCE: DLA opening position inventory stratification.

[a] Receipts of items occur throughout the year and are issued or disposed throughout the year, so we have multiplied this by 50 percent to account for the inflow and outflow of these items from the inventory.

[b] Average inventory on hand should be calculated as the midpoint between minimum and maximum insurance objective. Because that information was unavailable, we used the maximum insurance objective multiplied by 50 percent from the operating position instead.

The should-be level is primarily made up of the safety stock and the expected amount of cycle stock that would be on hand. For these numbers, we generally used the requirements inventory stratification in the opening position documentation from September 2011. For safety stock, the expected amount on hand would be the full safety-level requirements listed in the stratification. For cycle stock, the expected amount on hand is one-half of the order quantity; we thus use 50 percent of the procurement/order cycle requirements in the stratification. Items that are not forecastable (replenishment method code "R"), however, are not managed by order quantity and safety levels, but rather by maintaining a target level of stock known as the *insurance objective level*. For those items, the expected on-hand level would be halfway between the maximum level, given by the insurance objective, and the minimum level, which is not visible in the inventory stratification. In the absence of this minimum level, we have simply used the whole insurance objective for the should-be, thus overestimating the amount to some degree. But for true insurance items, the minimum should be close to the maximum, triggering replenishments upon reasonable levels of use.

The should-be level also includes additional inventory for which extra stock is kept for specific reasons. One category includes items for which an amount has been set aside as war reserve. For this, we use the on-hand amount because DLA does not buy inventory specifically for war reserve material but rather sometimes allocates existing inventory to this purpose.[4] The other category includes items for which DLA has

---

[4] Interview with DLA Headquarters personnel.

purchased inventory because of the dwindling availability of suppliers. For this, we use the requirements listed under diminishing manufacturing sources.

Computed Holding Cost Rate for Purchase Decisions
Having thus calculated the disposal value, storage costs, and inventory value, we combine these to get the estimated holding cost and compute a holding cost rate as a percentage of inventory value. For fiscal year 2011, the holding cost percentages are, on average, almost 23 percent, as shown in Table B.4.

These initial results of holding costs show large variation by supply chain. Note that in some supply chains, such as Aviation and Maritime, disposals are the dominant driver of holding costs, whereas in other supply chains, storage costs play as much of a, or even a larger, role. With additional years of data, analysts could determine whether using different holding costs by supply chain would be worthwhile for informing DLA's future purchases.

We note that we used fiscal year 2011 inventory stratification numbers in making our estimates of holding cost because those were available to us when we made the computations. We later received fiscal year 2013 inventory stratification numbers as well. However, those numbers reflected a transition in inventory stocking policies for items with infrequent but highly variable demand to new algorithms known as Peak

Table B.4
Estimated Holding Costs for Purchase Decisions, Fiscal Year 2011

Supply Chain	Avg. Value of Disposed Items ($ millions)	Storage Cost ($ millions)	Total Holding Cost ($ millions)	Should-Be Inventory ($ millions)	Holding Cost, Not Including Discount Rate (%)	Nominal Discount Rate (%)	Holding Cost (%)
Aviation	309.9	40.7	350.6	1,563.2	22.4	1	23.4
Construction and Equipment	27.9	28.0	55.9	158.0	35.4	1	36.4
Clothing and Textiles	22.4	32.8	55.2	703.4	7.9	1	8.9
Industrial Hardware	20.4	3.8	24.2	283.5	8.5	1	9.5
Land	58.8	51.8	110.6	353.9	31.2	1	32.2
Medical	3.5	2.4	5.9	13.0	45.2	1	46.2
Maritime	115.7	15.5	131.2	726.8	18.1	1	19.1
Total	651.2	175.0	826.2	3,801.8	21.7	1	22.7

NOTE: Total average value of disposed items and total holding costs include $92.6 million in disposals that were not labeled as being associated with a particular supply chain.

and NextGen. We opted to stay with the earlier inventory numbers because we deemed those values to better reflect the inventory policies in use at the time that the inventory excess, which now manifests as disposals, would have been generated.

We should also note that, even as inventory levels stabilize after the transition to Peak and NextGen, there will likely be variations in the holding costs seen from year to year due to the movement of disposal amounts, storage costs, and total should-be inventory. To prevent large swings in holding cost percentages from year to year, a multiple-year moving average could be used, as illustrated in this analysis.

Holding Costs for Disposal Decisions

Holding costs are also used to inform decisions on the disposal of excess inventory. While methods for computing retention amounts vary, they include a trade-off between the costs of continuing to hold the inventory and the costs of potentially later rebuying the items that were disposed.

Sunk Costs Should Not Be Included

The holding cost used for disposal decisions should not be the same as that used for purchasing decisions. The key difference lies in the timing of the commitment of money to inventory. With decisions on purchases, money has not yet been committed. Purchasing inventory opens DLA up to the risk of losing value due to obsolescence and eventual disposal. This is why disposal value is included as part of the holding cost, along with the storage cost.

In contrast, with decisions on disposals, the commitment of money lies in the past. Any risk of obsolescence has already been incurred. It is therefore a sunk cost and should not be included in the holding cost computation for disposal decisions. Consequently, the only holding cost that should be considered is the storage cost. The computation for the holding cost rate is thus:

$$\text{Holding cost rate for disposals} = \frac{\text{Storage cost}}{\text{Inventory value}}.$$

The storage cost is the same as that used in computing the holding cost for purchases, as shown in Table B.3. However, we use a different inventory value as the denominator of the holding cost rate computation.

Inventory Value Should Reflect Actual Holdings

To be consistent with how we have computed storage costs, the inventory value for the denominator of the holding cost rate for disposal decisions should likewise be based on the actual inventory holdings. We use operating position from the inventory stratification, which reflects the actual on-hand levels for computing inventory value. These are shown in Table B.5. These computations result in holding cost percentages for disposal decisions that are much smaller than those that should be used for purchase decisions.

Table B.5
Estimated Holding Costs for Disposal Decisions, Fiscal Year 2011

Supply Chain	Cost of Storage ($ millions)	On-Hand Inventory Value from Inventory Stratification ($ millions)	Holding Cost (%)
Aviation	40.7	5,910.9	0.69
Construction and Equipment	28.0	465.0	6.02
Clothing and Textiles	32.8	1,387.1	2.36
Industrial Hardware	3.8	931.7	0.41
Land	51.8	1,784.0	2.90
Medical	2.4	107.5	2.23
Maritime	15.5	2,333.6	0.66
Total	175.0	12,919.8	1.35

Economic Order Quantity Formulation

The use of the EOQ formula is pervasive in discussions of supply chain management and commonly used to set order quantities.[1] Given a set of assumptions, the EOQ formula determines the order quantity that minimizes total annual costs, which is the sum of the annual ordering cost (purchasing and receipt costs) and inventory holding cost. Using the notation in Table C.1, the total annual cost Z is given by the following equation,

$$Z = k\frac{\lambda}{Q} + hc\frac{Q}{2},$$

where the first term represents the annual ordering costs, and the second term represents the annual inventory holding costs. Note that this formulation assumes that other costs, including the purchase price per item and the receipt cost per item, do not change as the order quantity changes, and therefore are not included in the equation. More-complicated formulations exist for the cases in which this assumption does not hold, such as with discounts in the purchase price given for larger order quantities.

Table C.1
Economic Order Quantity Definitions

Parameter	Definition
Q	Order quantity
k	PR cost
λ	Annual demand rate
h	Holding cost rate (per dollar per year)
c	Unit purchase cost
Z	Total annual cost

[1]   Wilson, 1934; Harris, 1913.

The EOQ is the value Q^*, which minimizes the total annual cost Z. The solution is the EOQ formula, which is given by:

$$Q^* = \sqrt{\frac{2k\lambda}{hc}}.$$

Relationship of EOQ to Coverage Duration

The coverage duration T is the order quantity given in terms of days of supply, and it represents the expected number of days between orders. It can be determined by dividing the order quantity Q by the daily demand rate $\lambda / 365$. When the EOQ is used, the resulting coverage duration can be computed as:

$$T \text{ (in days)} = \frac{Q^*}{\dfrac{\lambda}{365}} = 365 * \sqrt{\frac{2k\lambda}{hc}} * \frac{1}{\lambda} = 365 * \sqrt{\frac{2k}{h}} \sqrt{\frac{1}{\lambda c}}.$$

Note that λc is the ADV. Thus, the optimal coverage duration is a nonlinear decreasing function of the ADV; as the ADV becomes higher, the coverage duration should be shorter in proportion to the square root of the relative change.

Computing the Average Inventory and Average Number of Purchase Requests When Using the EOQ

Limiting consideration to the cycle stock or the inventory on hand associated with the order quantity, which excludes expected on-hand inventory from safety stock, the average cycle stock inventory value for an item when the EOQ is used is determined by multiplying the unit cost c by the average inventory on hand, which is half the order quantity, or $Q / 2$. Combined with the EOQ formula, the average inventory value is:

$$\text{average inventory value} = c\frac{Q^*}{2} = \frac{c}{2}\sqrt{\frac{2k\lambda}{hc}} = \frac{1}{2}\sqrt{\frac{2k}{h}}\sqrt{\lambda c}.$$

The number of PRs per year of an item when the EOQ is used is determined by dividing the annual demand rate λ by the order quantity Q. Combined with the EOQ formula, the number of PRs per year is:

$$\text{PR per year} = \frac{\lambda}{Q^*} = \frac{\lambda}{\sqrt{\dfrac{2k\lambda}{hc}}} = \sqrt{\frac{h}{2k}}\sqrt{\lambda c}.$$

Note again that λc is the ADV. Thus, the average inventory value for an item is an increasing function of its ADV, increasing with the square root of the relative increase in demand rate and unit cost. The number of PRs per year for an item is also an increasing function of its ADV, increasing with the square root of the relative increase in demand and unit cost. Thus, for a given item with a set price, the average inventory value and number of PRs per year varies in proportion to the square root of annual demand. The price of the item shifts this curve up and down, again in proportion to the square root of the change in price.

Achieving a Target Inventory or Purchase Request Level

The order quantities computed using the EOQ will produce a total number of PRs per year, as well as a total average inventory level with respect to cycle stock. It may be that one or the other of these is higher than desired or feasible. For example, there may be a limit on the number of PRs that acquisition staff can handle in a year. Alternatively, there may be a restriction on the total inventory investment that is allowed.

To keep PR workload relatively feasible, DLA supply chains often use manual overrides, on an item-by-item basis, to force larger order quantities in order to bring down the total number of PRs per year. However, a more efficient way to accomplish this would be to adjust the cost parameters used in the EOQ calculation so that the order quantities of all the items are adjusted together, optimally balancing workload across items instead of changing order quantities for a subset of items. This concentrates workload reductions and inventory increases in a subset of items. Using manual overrides on an item-by-item basis may not be the most efficient way to work within constraints.

The EOQ formula gives the order quantity that best balances PR costs with inventory costs to minimize the sum of the two, according to the per-PR cost parameter k and the holding cost rate parameter h that are given. While the parameters that are used by DLA represent estimates of the actual costs incurred in processing a PR (k = \$441.55 / PR for non-LTC items) or in investing money into the holding of stock, these parameters can also be thought of as the knobs by which EOQ can be tuned to produce the desired or constrained PR workload or the desired inventory level.

Increasing k will cause each PR to be considered more expensive. This will lead the EOQ to produce larger order quantities so as to reduce the total number of PRs per year. These larger quantities will also mean that more inventory will be kept. Thus, the balance shifts away from more PRs and toward more inventory. The increase in the per-PR cost could also be thought of as adding an opportunity cost to each PR when workload constraints start producing a queue. Each PR not executed on time produces costs not accounted for in the EOQ model, including readiness costs.

Similarly, increasing h causes inventory to be considered more expensive. This will lead the EOQ to produce smaller order quantities so as to reduce the amount of inventory that is held. These smaller order quantities will also mean that orders will have to be placed more frequently, increasing the total number of PRs per year.

While either k or h could be adjusted to achieve either desired PR or inventory levels, it is easiest to think of adjusting the parameter that is more closely associated with the effect needed. Thus, if the goal is primarily to adjust the PR count, adjust the PR cost parameter k. Similarly, if the goal is primarily to adjust the inventory level, adjust the inventory cost parameter h.

It may also be easier to think of adjusting the parameters by multiplying them by a scaling factor. Recall that the EOQ formula for calculating the optimal order quantity Q is given by

$$Q^* = \sqrt{\frac{2k\lambda}{hc}}.$$

Suppose we multiply k by a scaling factor α so that the new PR cost is αk, and we multiply h by a factor of β so that the new holding cost rate is βh. The new order quantity, which we will call Q^{**}, would be:

$$Q^{**} = \sqrt{\frac{2\alpha k\lambda}{\beta hc}} = \sqrt{\frac{\alpha}{\beta}}\sqrt{\frac{2k\lambda}{hc}} = \sqrt{\frac{\alpha}{\beta}}Q^*.$$

Thus, the new order quantity is equal to the original order quantity, multiplied by a factor of

$$\sqrt{\frac{\alpha}{\beta}}.$$

The number of PRs per year with this new order quantity is

$$\frac{\lambda}{Q^{**}} = \frac{\lambda}{\sqrt{\dfrac{2\alpha k\lambda}{\beta hc}}} = \frac{1}{\sqrt{\dfrac{\alpha}{\beta}}} \frac{\lambda}{\sqrt{\dfrac{2\alpha k\lambda}{hc}}} = \sqrt{\frac{\beta}{\alpha}}\,\frac{\lambda}{Q^{*}}.$$

Thus, the new number of PRs is the old number of PRs, multiplied by a factor of

$$\sqrt{\frac{\beta}{\alpha}}.$$

The new average inventory value would be:

$$c\frac{Q^{**}}{2} = \frac{c}{2}\sqrt{\frac{2\alpha k\lambda}{\beta hc}} = \sqrt{\frac{\alpha}{\beta}}\left(\frac{c}{2}\sqrt{\frac{2k\lambda}{hc}}\right) = \sqrt{\frac{\alpha}{\beta}}\left(c\frac{Q^{*}}{2}\right).$$

Thus, the new average inventory value is equal to the old average inventory value, multiplied by a factor of

$$\sqrt{\frac{\alpha}{\beta}}.$$

These formulas make it simple to do a quick calculation of how much to increase the assumed PR cost in order to get the total number of PRs down by a desired level. For example, suppose DLA needed to decrease the number of PRs by 10 percent to reduce a backlog or to adjust to a change in the workforce size. To get the new number of PRs to be 90 percent of the old number of PRs would require setting

$$\sqrt{\frac{\beta}{\alpha}} = 0.9.$$

If we set $\beta = 1$, thus keeping the holding cost parameter constant, this would require setting scaling parameter for PR costs $\alpha = 1.23$. Thus, we would achieve a 10-percent reduction in PRs by scaling the PR cost parameter by a multiplicative factor of 1.23, changing it from the current \$441.55 to \$543.11.

As another example, suppose we needed to reduce the average inventory investment by 15 percent. We might wish to change the holding cost used in the EOQ calculation. To get the new inventory investment to be equal to 85 percent of the old inventory investment would require setting

$$\sqrt{\frac{\alpha}{\beta}} = 0.85.$$

If we set $\alpha = 1$, thus keeping the PR cost parameter constant, this would require setting the scaling parameter for holding costs $\beta = 1.38$. Thus, we would achieve a 15-percent reduction in inventory investment by scaling the holding cost parameter by a multiplicative factor of 1.38, changing it from the current 18 percent to 24.8 percent.

The formulas and examples apply not just for one item at a time, but for total PRs and total inventory across multiple items, as long as the same scaling factors are applied to each item. The total number of PRs per year, across all the items, would be:

$$\sum_i \frac{\lambda}{Q_i^{**}} = \sum_i \frac{\lambda}{\sqrt{\frac{2\alpha k \lambda_i}{\beta h c_i}}} = \left(\frac{1}{\sqrt{\frac{\alpha}{\beta}}}\right) \sum_i \frac{\lambda}{\sqrt{\frac{2\alpha k \lambda_i}{h c_i}}} = \left(\sqrt{\frac{\beta}{\alpha}}\right) \sum_i \frac{\lambda}{Q_i^{*}}.$$

Thus, the new total number of PRs is the old total number of PRs, multiplied by a factor of

$$\sqrt{\frac{\beta}{\alpha}}.$$

The new total average inventory value would be:

$$\sum_i c_i \frac{Q_i^{**}}{2} = \sum_i \frac{c_i}{2} \sqrt{\frac{2\alpha k \lambda_i}{\beta h c_i}} = \left(\sqrt{\frac{\alpha}{\beta}}\right) \sum_i \frac{c_i}{2} \sqrt{\frac{2 k \lambda_i}{h c_i}} = \left(\sqrt{\frac{\alpha}{\beta}}\right) \sum_i c_i \frac{Q_i^{*}}{2}.$$

Thus, the new total average inventory value is equal to the old total average inventory value, multiplied by a factor of

$$\sqrt{\frac{\alpha}{\beta}}.$$

Order Quantity and Safety Stock

This appendix examines the cost and performance implications of a variety of alternative rules for setting order quantities for DLA-managed forecastable items. Throughout this analysis, we model DLA's inventory policies with (Q, R) inventory models, where the key decision variables are the order quantity Q (how much to order) and the reorder point R (when to place an order), expressed in terms of inventory position (net inventory plus outstanding orders).

We begin with a review of EOQ and the inventory theory that serves as a basis for calculating the cost-optimal order quantity when demand is stochastic.[1] DLA's current order quantity policies are based on EOQ but include additional rules or constraints that make them suboptimal—as do manual overrides applied to them. We show that returning to EOQ can both lower inventory investment and reduce order workload. But we also point out that in certain cases, EOQ can be a very poor estimate of the optimal order quantity. Relying on EOQ without considering the relationship between order quantity and safety stock can be problematic, particularly for many DLA-managed items, which tend to have long lead times and highly variable demand. Using this insight, we then recommend an order quantity policy from a set of options put forward by DLA; discuss the effect of this policy on costs, workload, and agility; and point the way to further improvements.

Economic Order Quantity

For the purposes of this analysis, we represent a single item's inventory costs as the sum of its ordering costs and holding costs:[2]

---

[1] Wilson, 1934; Harris, 1913.

[2] Inventory costs can also include shortage or backorder costs. No official shortage cost—or even an estimate of what it might be—exists for DLA; correspondingly, shortage costs are not used by DLA for managing inventory. We ignore this cost and set service-level targets to ensure sufficient materiel availability, in line with DLA practices.

$$Z(Q,R) = \left(\frac{\lambda}{Q}\right)k + \left(\frac{Q}{2} + R - \lambda\tau\right)hc,$$

(D.1)

where λ is the expected annual demand, Q is the order quantity, k is the cost of placing an order, R is the reorder point, τ is the lead time in years, h is the annual holding cost rate expressed as a percentage of item cost, and c is the item's unit cost.

When demand is deterministic, the reorder point R can be set to exactly equal the lead-time demand ($\lambda\tau$) such that the next order arrives exactly as on-hand inventory drops to zero. This reduces the above cost equation to:

$$Z(Q,R) = \left(\frac{\lambda}{Q}\right)k + \left(\frac{Q}{2}\right)hc.$$

(D.2)

By taking the derivative of $Z(Q,R)$ with respect to Q (which determines the rate of change in cost with respect to Q), setting the result equal to zero, and solving for Q, we arrive at the optimal (cost-minimizing) order quantity under deterministic demand, the familiar EOQ:

$$EOQ = \sqrt{\frac{2k\lambda}{hc}}.$$

(D.3)

Stochastic Demand and Materiel Availability

We assume that lead-time demand in excess of the stock available at the time an order is placed is backordered and then fulfilled when the order arrives.

When demand is stochastic, inventory managers typically use forecasting to estimate lead-time demand and to set the reorder point. These forecasts will, by definition, err because demand is stochastic, meaning that actual demand may be greater or lower than expected. Greater-than-expected demand can be fulfilled at a desired level of confidence from *safety stock*—the extra inventory held to ensure that materiel availability targets are met in the face of unexpectedly large lead-time demand. Safety levels are set by simply raising the reorder point R above the expected lead-time demand $\lambda\tau$, which, all else equal, has the effect of increasing the average on-hand inventory level. Materiel availability (or fill rate) is expressed as the proportion of annual demand immediately satisfied from on-hand inventory. The expected number of backorders per order cycle can be approximated as:

$$n(r) = \int_R^\infty (x - R)f(x)\,dx,$$

(D.4)

where $f(x)$ is the probability distribution function of lead-time demand, x, and the number of order cycles per year is $\lambda \div Q$. Therefore, the expected materiel availability for a given Q and R can be approximated as:

$$\beta = 1 - \frac{\int_{R}^{\infty} (x - R) f(x) \, dx}{Q}.$$

(D.5)

Based on the equation above, we can see that the fill rate β can be increased by either raising the reorder point R or increasing the order quantity Q. Raising R can be achieved by carrying more safety stock, creating a larger buffer against stockouts. Ordering larger quantities decreases the frequency with which inventory levels sink low enough to be at risk of a stockout. Figure D.1 helps demonstrate these relationships.

In fact, a given fill rate can be achieved using any number of combinations of Q and R. The curve in the Figure D.2 highlights the combinations of Q and R that together achieve a target fill rate of 90 percent for the listed mean and standard deviation of lead-time demand, under the assumption that lead-time demand is normally distributed. EOQ represents just one point along this curve.

When demand is stochastic, safety stock is typically used to achieve a desired fill rate. As discussed, this raises the reorder point R above $\lambda\tau$, which means that the simpler cost equation (Equation D.2) no longer applies and that EOQ no longer minimizes total cost. In this case, we must use Equation D.1 to calculate total cost, and

Figure D.1
Materiel Availability Can Be Increased by Increasing Either Q or R

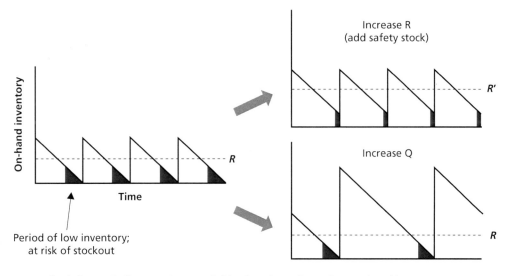

NOTE: Shaded areas indicate an increased risk of stockout due to low on-hand inventory.

Figure D.2
A Target Fill Rate Can Be Achieved with a Variety of Combinations of Q and R

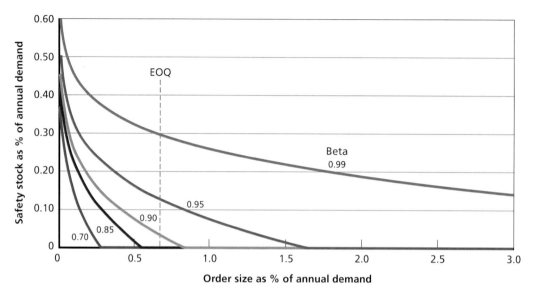

NOTES: Curves calculated for example item, where λ = 1,000 units per year, coefficient of variance (CV) = 1, τ = 0.5 years, k = $441 per order, c = $10 per unit, h = 18 percent, and lead-time demand forecast errors are normally distributed.
RAND RR822-D.2

we calculate the optimal order quantity Q^* using the following formula, where $F(R)$ represents the cumulative distribution function for lead-time demand:[3]

$$Q^* = \frac{n(R)}{1-F(R)} + \sqrt{\frac{2k\lambda}{hc} + \left(\frac{n(R)}{1-F(R)}\right)^2}.$$

(D.6)

Note that the formula for Q^* depends on R and must be solved for iteratively with $n(r) = Q(1 - \beta)$. Because Equation D.6 does not have a closed-form solution, inventory managers often use $Q = \text{EOQ}$ instead, even with stochastic demand, assuming EOQ to be a sufficiently good approximation of the optimal order quantity.

While this simplifying assumption works reasonably well most of the time, in certain cases, it does not. To illustrate, Figure D.3 shows two curves for an actual DLA-managed item. The left curve identifies the combinations of Q and R that together achieve the desired fill rate of 90 percent, with Q on the x-axis and R on the y-axis. The

---

[3] This formula is derived following the same approach as the derivation of EOQ—that is, taking the derivative of the cost equation, in this case Equation D.1, with respect to cost, setting the result equal to zero, and solving for Q. For an example of this derivation, see J. A. Muckstadt and A. Sapra, *Principles of Inventory Management: When You Are Down to Four, Order More*, New York: Springer Science + Business Media, 2010.

Figure D.3
Total Costs Under EOQ and Q^* for Item 014634340

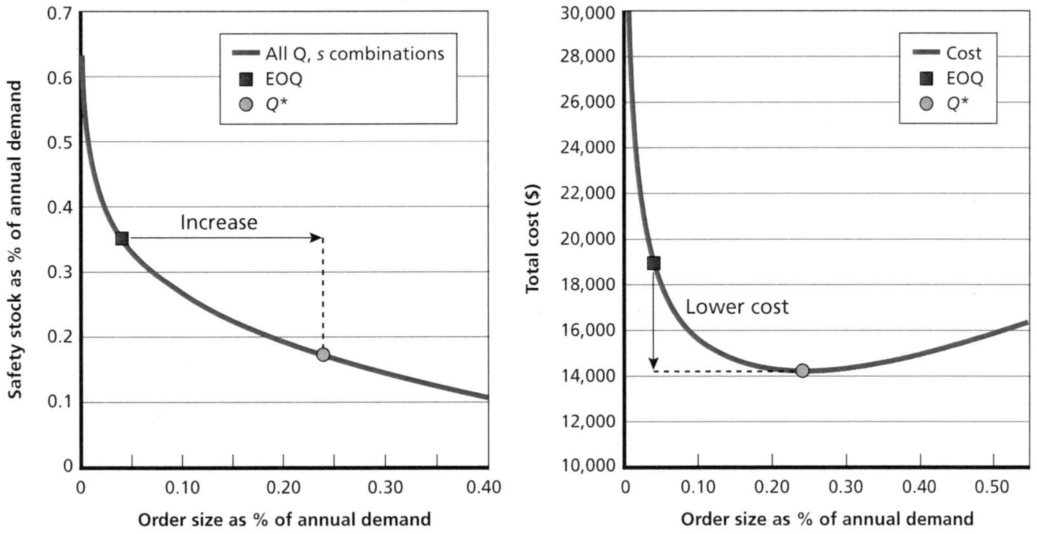

NOTES: Curves calculated for Item 014634340, where λ = 1,638 units per year, coefficient of variance (CV) = 0.84, τ = 9 months, k = $441 per order, c = $1,631 per unit, h = 18 percent, and lead-time demand forecast errors are normally distributed.
RAND RR822-D.3

right curve shows total cost as a function of Q (where R is set to achieve the desired fill rate given the selection of Q). As can be seen, while EOQ (red squares) minimizes ordering costs and cycle stock costs, it does not take into account the effect of Q on the required safety stock. This additional cost shifts the total cost curve and makes using Q = EOQ quite costly compared with $Q = Q^*$.

When Is EOQ a Poor Approximation of Optimal?

The preceding example motivates the following question: Under what conditions is EOQ a poor approximation of the optimal order quantity in terms of total cost? To better understand the cost implications of using EOQ instead of Q^*, we conducted an experiment in which we generated a variety of combinations of key item attributes. Then for each combination, we calculated both EOQ and Q^* and compared total costs under each quantity. Finally, we searched for a relationship between the ratio of their costs and the values of the candidate attributes. Table D.1 lists the combinations of attributes and their candidate values.

These attributes can be combined to represent 9,000 unique hypothetical items. Figure D.4 shows the subset of these combinations for which the target fill rate is 90 percent. The curve plots the ratio of EOQ costs to Q^* costs as a function of the ratio

Table D.1
Attributes and Values for Order Size Cost Comparison

Order Cost	Unit Price	Annual Demand	Coefficient of Variance	Lead Time (months)	Fill Rate (%)
$21	$1	10	0.1	1	70
$441	$10	100	0.25	3	85
	$100	1,000	0.5	6	90
	$1,000	10,000	1	12	95
	$10,000	100,000	2	24	99
	$100,000		4		

Figure D.4
Ratio of Economic Order Quantity Cost to Q^* Cost as a Function of EOQ/σ_r

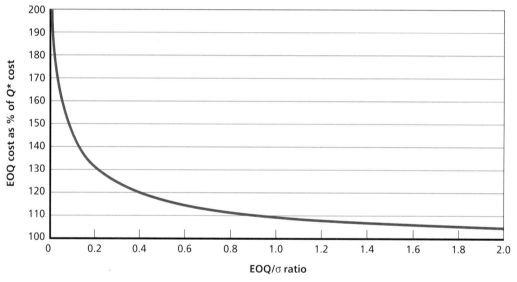

of EOQ to the standard deviation of lead-time demand. The curve demonstrates that if EOQ is small or if lead-time demand is highly variable, using EOQ instead of Q^* can lead to significantly higher costs. EOQ tends to be small when annual demand λ is low, ordering costs k are low, or unit price is high (c – the holding cost rate h is assumed to be fixed). Examining the formulas for EOQ and Q^*, we conclude that EOQ may be a poor approximation of Q^* for items with any combination of high lead-time demand variability, high unit price, low ordering costs, and low annual demand. The same relationship exists for other fill rate goals.

Applicability of EOQ to DLA Items

EOQ is widely used in inventory management textbooks and in the field as a reasonable approximation of the optimal order quantity. However, understanding the assumptions of such an approximation is critical. DLA chiefly manages spare parts, which are frequently characterized by sparse and highly stochastic demand; furthermore, lead times for DLA items often span many months. EOQ may closely approximate the optimal order quantity for items with low demand variability and short lead times, but this approximation can break down when applied to DLA items.

Special Implications for Items Placed on Long-Term Contracts

As discussed above, the ratio of EOQ to the standard deviation of lead-time demand can serve as an indicator of the quality of EOQ as an approximation of Q^*. While targeting a particular ratio is probably misguided, this ratio can demonstrate the problems with blindly using EOQ as the order quantity.

Placing an item on LTC (or *outline agreement*) has many advantages, including lowering ordering costs and shortening administrative lead times. Under DLA's assumptions, when an item is placed on LTC, order costs reduce from around $441 per order to just $21 per order. All else equal, this results in a reduction of EOQ by a factor of about 4.6, or about 78 percent.

To maintain the same ratio (i.e., for EOQ to be at least as good an approximation of Q^* once the item is placed on LTC), the standard deviation of lead-time demand would have to also be reduced by a factor of 4.6. This value is calculated as $\sigma\sqrt{\tau}$, where σ represents the standard deviation of demand per unit time and τ represents the lead time. Because σ is unlikely to be affected by the item's contract status, the only way to achieve a reduction is to reduce τ. Because standard deviation of lead-time demand is a function of the square root of τ, τ must be reduced by a factor of 4.6^2, or 21 (a 95-percent reduction) in order to achieve the necessary reduction in standard deviation of lead-time demand. While lead time (particularly ALT and lead time for items ordered in large quantities but delivered in small batches) does tend to become shorter when an item is placed on LTC, a reduction in total lead time of 95 percent is highly unrealistic, meaning that EOQ is likely to be a less accurate approximation for Q^* once the item is placed on LTC.

Current DLA Order Quantity Policies

As described elsewhere in this report, DLA currently uses its coverage duration (CovDur) rules to set order quantities for forecastable items. Under CovDur, DLA sets order quantities in terms of *days of supply*—the number of days of demand that the order quantity can satisfy. CovDur maps ranges of ADVs to days of supply, and is based on

EOQ, but uses an estimated holding cost h of 36 percent rather than DLA's estimated value of 18 percent. This high holding cost was intended to constrain inventory investment to the existing level when DLA adopted the use of EOQs; however, while higher holding costs produce smaller order quantities and reduced cycle stock investment, the smaller order quantities may increase safety stock requirements such that total inventory investment actually increases. If actual holding costs are closer to 18 percent than 36 percent, then CovDur produces unnecessarily small order quantities.

DLA further modifies CovDur on a case-by-case basis to arrive at the actual order quantities (System CovDur or MPSCOVDUR in DLA's Enterprise Business System, EBS). Figure D.5 depicts how actual order quantities compare with EOQ, using an 18-percent holding cost percentage, as a function of ADV for non-LTC items.

An examination of the optimal order quantity formula introduced earlier reveals that the optimal order quantity is always at least as large as EOQ. As can be seen in the figure, the CovDur (the red curve) is frequently set to less than EOQ (the green curve); this implies that even if no safety stock is required, the CovDur order quantity would result in excess orders and unnecessary costs. Furthermore, the striation of actual order quantities (the blue dots) for any given ADV reflects widespread inconsistency in the application of DLA's own order quantity policies. In some cases, quantities larger than EOQs may reduce the need for safety stock and lower total cost; however, the actual order quantity should always be at least as large as EOQ. The large number of dots that appear below the EOQ curve suggest ample opportunity for reducing costs and order

Figure D.5
Comparison of Economic Order Quantity, Coverage Duration, and Actual Order Quantities for Non-LTC Items

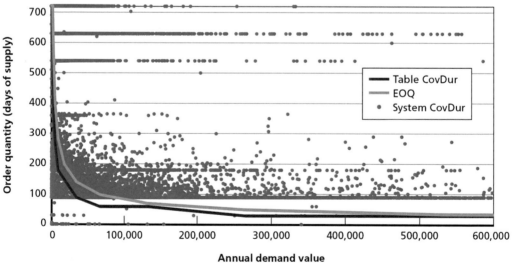

workload (and, even if CovDur's 36-percent holding cost is accurate, there are still many instances of actual order quantities that are less than even what CovDur would suggest). Note also that there is a 90-day minimum in effect for most items.

Alternative Order Quantity Analysis

The following analysis examines the effect on inventory investment, order workload, and total cost of each of several alternative order quantity rules. The goal of this analysis is to help illuminate the implications of order quantity decisions on these key metrics, and to provide context for making changes to the order quantity policies used in practice.

Practical Bounds for Order Quantities

For a variety of reasons, DLA imposes constraints on its order sizes. For example, because forecasting demand beyond two years will tend to have large errors, DoD policy imposes an upper bound on order quantities of no more than 720 days' (two years') worth of demand.[4] This reduces the obsolescence risk associated with placing very large orders (e.g., in exchange for quantity discounts). In addition, DoD policy imposes a lower bound on order quantities of the lesser of one month's demand or the expected demand over the ALT. For the purposes of the subsequent analysis, we will assume that order quantities can be no less than seven days of supply and no more than 720 days of supply, with the exception of EOQ, for which we do not impose a lower bound.

Fill Rates and Safety Stock

DLA uses a variety of performance targets at the wholesale and retail echelons of its inventory system to determine how to set inventory levels. For simplicity, we assume that DLA seeks to achieve a 90-percent fill rate for each of the items it stocks (rather than a systemwide average or average by supply chain). The rest of this analysis will make comparisons between alternative order policies assuming that the service-level goal is set for each item independently. (This contrasts with DLA's approach of optimizing safety stock investment across items, which reduces total safety stock investment by using higher fill rates for cheaper items with relatively constant demand and lower fill rates for expensive items with more variable demand.) We also assume that excess demand is backordered, that forecast errors for lead-time demand are stationary and normally distributed (though we will also discuss the implications of other distributions), and that lead times are deterministic.

---

[4]   U.S. Department of Defense, *DoD Supply Chain Materiel Management Procedures: Demand and Supply Planning*, DoD Manual 4140.01, Vol. 2, February 2014a.

Data

The analysis covers all forecastable items managed by DLA. Data on these items' ordering costs, unit prices, means and standard deviations of demand, and lead times were collected from EBS, DLA's enterprise resource planning system.

Alternative Order Sizes

Table D.2 lists a series of alternative order size rules, for both LTC and non-LTC items. The first, System, represents the actual order quantities in use as of November 2013. CovDur represents the order quantities as determined by the DLA CovDur Table and EBS rules (without the overrides that produce System CovDur values). EOQ and EOQ30 represent the traditional economic order quantity and an alternative that stipulates that the order quantity cannot be less than 30 days of supply, respectively. During the course of this study, DLA made a decision to convert to using EOQ with an 18-percent holding cost, coupled with standardized upper and lower bound constraints (as opposed to item-by-item manual overrides). Alternatives I, II, and III represent the three alternative rules considered by DLA. Finally, the true optimal order quantity, Q^*, is included for comparison purposes.

Table D.2
Alternative Rules for Order Quantities

Q	Long-Term Contract Items			Non–Long-Term Contract Items		
	Min	Default	Max	Min	Default	Max
System	N/A	MPSCOVDUR	N/A	N/A	MPSCOVDUR	N/A
CovDur	N/A	30 days if ADV > 100K; 90 days otherwise	N/A	N/A	Table CovDur	N/A
EOQ	N/A	EOQ	720 days	N/A	EOQ	720 days
EOQ30	30 days	EOQ	720 days	30 days	EOQ	720 days
Alternative I	7 days	EOQ	720 days	ALT	EOQ	720 days
Alternative II	30 days if ADV > 100K; 90 days otherwise	EOQ	720 days	Max(90, ALT)	EOQ	720 days
Alternative III	Non-Aviation: 30 days if ADV > 100K; 90 days otherwise. Aviation: Max(90, ALT)	EOQ	720 days	Max(90, ALT)	EOQ	720 days
Q^*	N/A	Q^*	720 days	N/A	Q^*	720 days

NOTE: N/A is not applicable.

Inventory Investment and Purchase Requests: LTC Items

Figure D.6 depicts the total inventory investment (both in cycle stock and safety stock) and the PR workload for each of the alternatives described above. Relative to the current state (System), all of the alternatives increase the number of orders. However, most of the labor involved with issuing a PR is automated for LTC items, and thus this does not represent an increase in the burden on DLA supply planners.

In keeping with the preceding discussion, we can see that setting order quantities equal to EOQ results in a dramatic reduction in cycle stock, but also a steep increase in safety stock (recall that EOQ does not take safety stock costs into account). This safety stock investment is necessary to compensate for the fact that EOQ's relatively small order quantities would otherwise significantly increase the probability of a stockout. This effect can be reduced by raising the lower bound on EOQ to 30 days of supply (EOQ30). This results in a slight increase in cycle stock but allows for a significant reduction in safety stock.

Note that DoD policy constrains order quantities to be at least as large as the lesser of expected demand over one month or over the ALT, and that DoD limits safety stock investments to a maximum of three standard deviations of lead-time demand. Therefore, any DLA implementation of EOQ would necessarily include additional constraints. The inventory investment shown here for unconstrained EOQ is somewhat unrealistic; EOQ30 is probably a better representation of the lowest reasonable order quantity levels for DLA at present.

Figure D.6
Inventory Investment and Purchase Requests for LTC Items, by Order Quantity

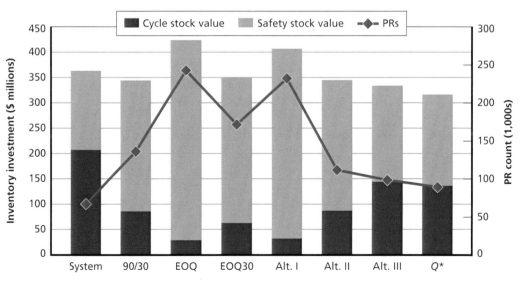

As with unconstrained EOQ, Alternative I, with its very low minimum order quantity of seven days, also produces large safety stock investments.

Alternative II closely matches the performance of the Table CovDur entry, which is not surprising given the similarity of the policies. Of all the options, Alternative III comes closest to matching the performance of the true optimal quantity. Alternative III uses EOQ by default but introduces relatively high minimums, which increases order quantities above EOQ for many items. This mimics the effect achieved by Q^*, in which order quantities are increased to reduce safety stock investment in exchange for comparably smaller cycle stock investment. While Alternative III performs well compared with the other options examined here, it still does not match the performance of the optimum quantity, resulting in about 6 percent more in inventory investment and about 10 percent more purchase requests.

Inventory Investment and Purchase Requests: Non-LTC Items

Figure D.7 shows the same metrics as Figure D.6, but now we evaluate the results for non-LTC items. Here we see that Table CovDur actually produces less cycle stock investment and more safety stock investment than EOQ for a lower total expected on-hand inventory. However, it requires more orders and procurement workload. This is consistent with its use of a 36-percent holding cost rate versus 18 percent. While Table CovDur results in less overall inventory investment, the cost of these smaller order quantities manifests itself in the order workload, which is of concern because placing orders for items not on LTC can involve a labor-intensive solicitation process.

Figure D.7
Inventory Investment and Purchase Requests for Non-LTC Items, by Order Quantity

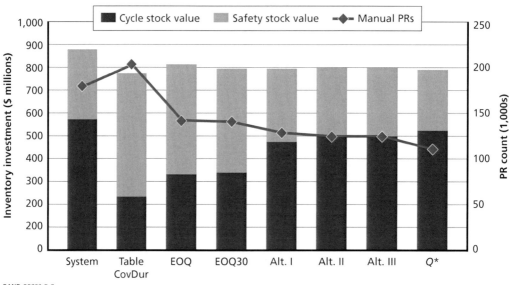

The other alternatives produce nearly identical total inventory investment, though predictably, EOQ and EOQ30 have smaller order sizes, more orders, less cycle stock, and more safety stock than the other alternatives. Alternatives I, II, and III all have very similar rules for non-LTC items and produce very comparable results. Again, the optimal quantity, Q^*, still offers improved performance over the other alternatives; in this case, Q^* better allocates inventory investment across cycle stock and safety stock, resulting in roughly identical total inventory investment but fewer orders.

Total Cost

Figure D.8 presents the effect of each order quantity alternative on total cost and order workload. As we would expect, the optimal order quantity produces the lowest total cost and the fewest orders—both in terms of total orders and manual (non-LTC) orders.

Table CovDur outperforms the current order quantities, suggesting that DLA could improve its performance simply by more closely adhering to its current policies. However, Table CovDur still leaves ample room for improvement. While ordering the true optimal order quantity may not be feasible given the constraints of DLA's inventory management systems, Alternative III comes close to achieving the optimal performance in terms of total cost and order workload (and Alternative II is a close second). Thus, with relatively simple changes to its ordering policies, DLA could significantly improve over the current state and come within about 5 percent of the optimal total cost.

Figure D.8
Total Cost and Purchase Requests, by Order Quantity

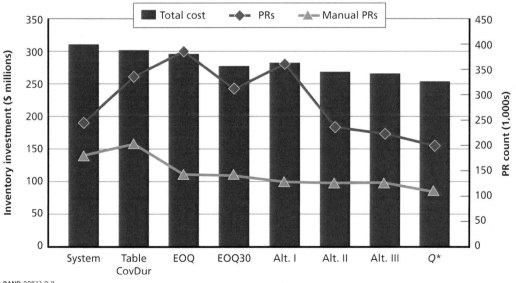

Recommended Approach to Order Quantity

The variety of items managed by DLA and the complexity of the systems used to manage these items suggest an approach to order quantity that is evolutionary and seeks to continuously improve performance. Current order quantities reflect an accumulation of one-off changes and constraints that together produce inventory investment and order workloads that perform significantly worse than the optimal. Rather than immediately switching to ordering the optimal quantity—something that itself may be difficult to define without making overly simplifying assumptions—DLA should seek to iteratively improve the rules it uses to make order quantity decisions.

For example, earlier we noted that the optimal order quantity is always at least as large as EOQ. One relatively simple step that DLA could take would be to determine the appropriate holding cost rate to use in its calculations of order quantity. As discussed, DLA's CovDur equation is based on EOQ using a 36-percent holding cost rate even though DLA typically assumes an 18-percent holding cost rate. If the 18-percent figure is correct, CovDur values may be too small, resulting in more orders and more safety stock than necessary. Verifying this value, and updating if necessary, would ensure that all orders are at least as large as what is suggested by EOQ (except in special situations, such as when a part is being phased out).

A more significant change would be to move toward order quantity policies similar to those described under Alternative III. Alternative III achieved inventory investment, order workload, and total cost performance comparable to the theoretical optimal order quantity Q^*, and of all the order quantities examined here, it performed the best.

However, it is important to note that even Alternative III produces order quantities that are significantly suboptimal in some cases. Overall, Alternative III achieves total costs that are about 5 percent worse than the optimal order quantity. This shortfall is driven primarily by a small subset of items for which Alternative III performs particularly poorly. For one such item, NIIN 015782781 ("Vane Segment"), a non-LTC item in the Aviation supply chain, Alternative III suggests an order quantity of 3,582 units, while the optimal order quantity is only 530 units. This results in a significantly larger cycle stock investment and increases total costs by nearly three times. There are hundreds of similar examples, but out of a total of more than 150,000 items, this represents a small minority of items. Nevertheless, if DLA adopts Alternative III, as it is currently in the process of doing, it would be best served to view these order quantity rules as an interim step toward further improvement.

Additional Discussion

Relationship Between Order Size and Lead Time

As we note in Appendix G, order quantity was one of the many factors found to be linked to lead times. While models may, for simplicity, assume that order quantity and lead time are independent, they are of course linked in the real world. The analysis presented in Appendix G found only a weak correlation between large orders and longer administrative and production lead times; however, it is important to keep in mind that extreme changes in order quantities may produce significant changes in lead times, which could affect the amount of safety stock required to achieve a desired fill rate over the lead time.

Simple Approximations for Optimal Order Quantity

A number of simple, noniterative approximations for the optimal order quantity Q^* have been put forward in the inventory management literature. For example, Brown introduced an approximation for Q^* that he called the *stock optimal quantity*:[5]

$$SOQ = \delta\sigma + \sqrt{(\delta\sigma)^2 + \frac{2k\lambda}{hc}},$$

(D.7)

where δ is set for a particular fill rate (Brown used $\delta = 0.559$ for 90-percent fill rate). This formula reduces to EOQ when the standard deviation of lead-time demand is very small, but produces a larger order quantity when lead-time demand is highly variable. Mabin developed a similar approach but used least squares regression to find the relationship between Q^*/σ and the corresponding reorder point R, fitting to a logarithmic function $R = \alpha - \delta\log(Q^*/\sigma)$.[6] She arrived at the same function as Brown (Equation D.7) but calculated a variety of different values for δ corresponding to different service levels (e.g., $\delta = 0.713$ for 80 percent, $\delta = 0.559$ for 90 percent, and $\delta = 0.474$ for 95 percent). These approximations are easy to calculate and often produce estimates for the optimal order quantity that are very close to the true value. However, care must be taken when applying these formulas; generally speaking, they are customized to a particular lead-time demand distribution (e.g., normal) and target fill rate (e.g., 90 percent).

Implications of Inventory Modeling Assumptions
Multiple Outstanding Orders

In the preceding analysis, we have used a simplified model for the expected percentage of demand that is unfilled from on-hand inventory (i.e., backordered). This model (Equation D.5) is relatively easy to work with, particularly when one assumes that forecast errors for lead-time demand are normally distributed. One of the key assumptions of this model is that there can be no more than one order outstanding at a time.

---

[5] Robert G. Brown, *Decision Rules for Inventory Management*, New York: Holt, Rinehart and Winston, 1967.

[6] Victoria J. Mabin, "A Practical Near-Optimal Order Quantity Method," *Engineering Costs and Production Economics*, Vol. 15, 1988, pp. 381–386.

Equation D.8 presents a more accurate (though still approximate) model for normally distributed demand that relaxes this assumption:

$$\beta = 1 - \frac{\int_{R}^{\infty}(x-R)f(x)dx - \int_{R+Q}^{\infty}(x-R)f(x)dx}{Q}.$$

(D.8)

In this formulation, the reorder point R is set in terms of inventory position rather than on-hand inventory.

While Equation D.8 is more accurate, it is also more difficult to use for trying to find the cost-optimal values for Q and R subject to a fill rate constraint. This is because the resulting cost function is not necessarily convex, unless one assumes that Q is more than two standard deviations larger than expected lead-time demand, which reduces the second integral in Equation D.8 to nearly zero. (This of course makes having more than one outstanding order unlikely, which diminishes the usefulness of this more accurate model.) However, approximately optimal values can be found using this model with enumerative search.

Two questions come to mind regarding the more accurate model: Does using Equation D.8 to find Q^* produce values for Q^* significantly different from the approximate model (Equation D.5), and does using Equation D.8 to select R affect the potential cost savings of Q^* relative to EOQ?

To help address these questions, we calculated EOQ and Q^*, as well as R and total costs, using both models, and then compared the results.

While there are differences in the values for Q^*, R, and total cost, the overall results are directionally the same. For both models, using Q^* yields total costs that are as low or lower than those under EOQ (or any other order quantity); Q^* is always at least as large as EOQ; and when lead-time demand is highly variable, Q^* can be significantly larger than EOQ. However, the simplified model overstates the effect: The increase in quantity between EOQ and Q^* tends to be less when using the more accurate model than when using the simplified model.

Of the approximately 150,000 items analyzed, approximately 39 percent have optimal order quantities that are larger than EOQ for one or both models. In 70 percent of these cases, the two models produced identical values for Q^*. For items where the estimates of Q^* did not match, the simplified model produced larger estimates of Q^* 85 percent of the time, while the more accurate model produced larger estimates of Q^* 15 percent of the time.

Figure D.9 shows the size of Q^* in terms of EOQ for each model as a function of the ratio of EOQ to the standard deviation of lead-time demand (σ_r). Though the models do produce different estimates of Q^*, particularly for items with small ratios of EOQ to σ_r, the general relationship between Q^* and EOQ as a function of the

Figure D.9
Q^*, in Terms of EOQ, as a Function of EOQ/σ Ratio

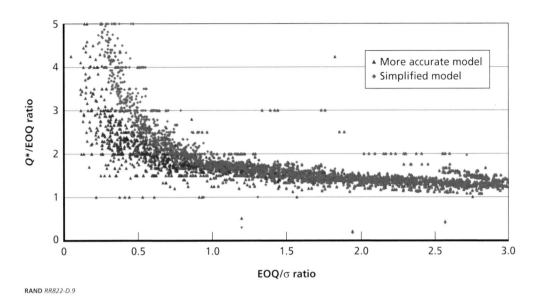

RAND *RR822-D.9*

ratio of EOQ to σ_τ is consistent; that is, both models produce larger Q^* estimates (in terms of EOQ) when the ratio of EOQ to σ_τ is small.

However, while the differences in each model's estimate of Q^* are relatively minor, there are more-notable differences in each model's effect on total costs. As shown in Figure D.10, for the more accurate model, the potential cost savings of using the optimal order quantity instead of EOQ increase as the ratio of EOQ to σ_τ diminishes—but only to a point. Recall that the more accurate model allows for more than one order to be outstanding at a time. This means that during the lead time for one order, net inventory can be replenished by the arrival of prior orders. This reduces the need for safety stock, which leads to lower total costs, and for some items, diminishes the potential cost savings of using Q^* rather than EOQ.

However, it is important to note that this is true only if one uses the same model to set safety stock levels that is used to set order quantities. In other words, if the logic used to set safety stock levels relies on the more commonly used simplified model, one should also use the simplified model to solve for Q^*, and vice versa.

Alternative Lead-Time Demand Distributions
Throughout this analysis, we have assumed that forecast errors for demand during the lead time are normally distributed. This is a common assumption in inventory modeling because the normal distribution is easy to work with and it is reasonably accurate for items with low variability. However, we also wanted to explore how our findings might change if an alternative distribution is used.

Figure D.10
EOQ Costs, in Terms of $Q*$ Costs, as a Function of EOQ/σ Ratio

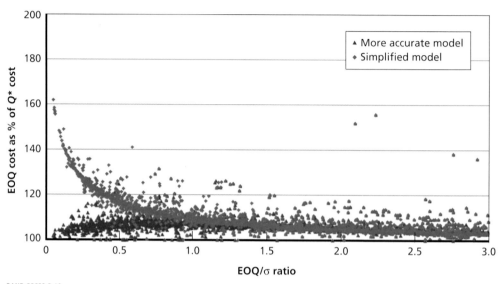

The Laplace and negative binomial distributions are two common alternatives for modeling lead-time demand for military spare parts.

Presutti and Trepp recognized that using the Laplace distribution to model lead-time demand can significantly simplify calculations of various performance measures of interest.[7] The Laplace distribution is also often considered preferable because it has *fatter tails* than the normal distribution; that is, it assigns higher probability to very large lead-time demands than does the normal distribution, which may better represent actual demand patterns. As with the normal distribution, optimization of Q and R using the Laplace distribution requires numerical search.

Deemer, Kaplan, and Kruse used a negative binomial distribution to model lead-time demand.[8] The negative binomial distribution also has fatter tails and has the advantage of assigning no probability to negative demands (not necessarily true for normal or Laplace distributions), but it is a discrete distribution, making it computationally more difficult to work with. Optimizing Q and R using a negative binomial distribution requires enumerative search.

Figure D.11 shows the upper quartile of the cumulative distribution functions of each of these three distributions fit to the same random variable, x. We can see that

[7]   Victor J. Presutti, Jr., and Richard C. Trepp, *More Ado About Economic Order Quantity (EOQ)*, Operations Research Office, Headquarters, Air Force Logistics Command, 1970.

[8]   R. L. Deemer, A. J. Kaplan, and W. K. Kruse, *Application of Negative Binomial Probability to Inventory Control*, AMC Inventory Research Office, Institute of Logistics Research ALMC, Philadelphia, Pa., 1974.

Figure D.11
Cumulative Probability for Normal, Laplace, and Negative Binomial Distributions

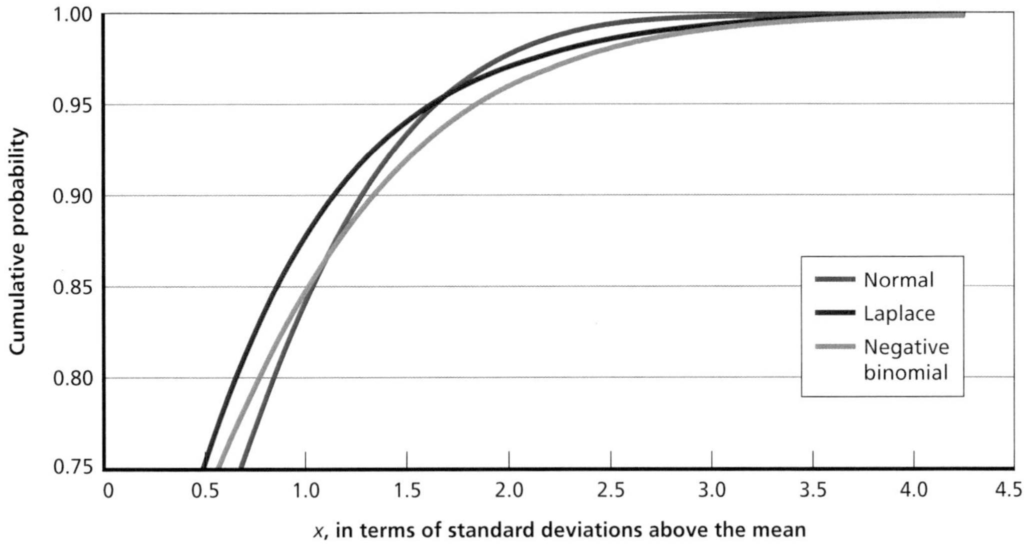

the Laplace distribution does indeed have more probability at extreme values (i.e., the Laplace curve is below the Normal curve) but that this occurs only when x is greater than about 1.69 standard deviations above the mean, or once cumulative probability exceeds about 95 percent for both distributions. The negative binomial distribution also places significantly more probability on extreme values but exceeds the normal distribution when x is beyond only 1.12 standard deviations above the mean, or when the cumulative probability for both distributions exceeds about 87 percent. This has implications for our analysis because our fill rate goals are set to 90 percent. While the Laplace distribution's fatter tails are often thought to make it better suited to modeling highly variable lead-time demand, all else equal, it actually requires less safety stock to achieve fill rate targets *below* about 95 percent than does the normal distribution. By contrast, the negative binomial distribution, all else equal, generally requires more safety stock to achieve fill rate targets above 87 percent (e.g., our customary 90-percent fill goal).

An example helps illustrate this effect. Figure D.12 shows the required safety stock as a function of order quantity if lead-time demand is modeled using each of the three distributions for Item 015350972.[9] As expected, for a 90-percent fill rate

---

[9]   Note that here we use the more accurate model (Equation D.8) for the expected fill rate for normal demand because the models for Laplace and negative binomial distributed demand follow the same form. This is by necessity; using the simplified model with Laplace distribution produces an integral that cannot be evaluated in closed form.

Figure D.12
Safety Stock as a Function of Order Quantity Using Three Alternative Lead-Time Demand Distributions for Item 015350972

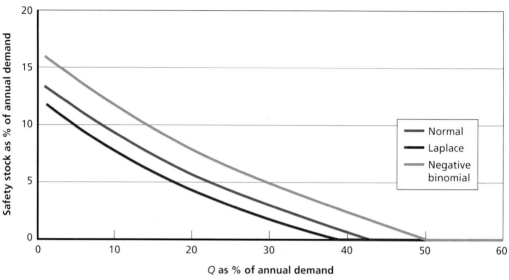

goal, Laplace demand requires the least safety stock for any given Q, negative binomial demand requires the most safety stock for any given Q, and normal demand falls somewhere in between.

However, it is important to note that there is significant overlap between the values produced by each distribution, and in many cases, the distributions recommend the same values for Q^* and R. To help demonstrate this, we calculated R for both EOQ and Q^* using each of the three lead-time distributions for a 10-percent sample of the full item population (approximately 15,000 items). We then analyzed how the value for Q^* compared with EOQ for each of the distributions. Figure D.13 shows a histogram depicting the frequency with which Q^* exceeds EOQ by a given amount for each of the three distributions.

As Figure D.13 shows, the three distributions produce relatively consistent values for Q^*, but there are some differences. In general, using the Laplace distribution tends to produce more Q^* values that are between 0 percent and 25 percent bigger than EOQ. Similarly, the normal distribution tends to produce more Q^* values between 30 percent and 60 percent bigger than EOQ. And finally, the negative binomial distribution tends to produce more Q^* values that are between 60 percent and 80 percent bigger than EOQ. Thus, the extent to which Q^* exceeds EOQ depends to a degree on the distribution used to model forecast errors over the lead time. While the results will vary from item to item, those items with demands that more closely follow a Laplace distribution will have Q^* values that tend to be closer to EOQ, and those

Figure D.13
Frequency of Increase in Q^* Relative to EOQ for Alternative Demand Distributions

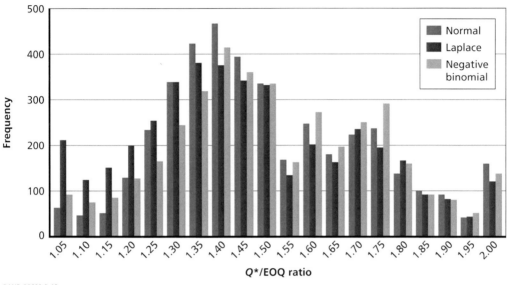

RAND *RR822-D.13*

with demands that follow a negative binomial distribution will have Q^* values that tend to be further from (i.e., larger than) EOQ, compared with Q^* estimates based on the normal distribution. A query of DLA's EBS indicated that as of February 2014, a majority of items are modeled using the negative binomial distribution. This suggests that, all else equal, our findings, based on the normal distribution, may slightly underestimate the real effect for these items.

Private-Sector Practices to Reduce Lead Time

Reducing lead times is a goal pursued not only by DLA but by industry. Because some strategies used in industry may apply to DLA, we sought to identify those that companies use to reduce lead time. These strategies can serve as a point of comparison for DLA practices and help identify opportunities for improvement, while also considering unique DLA and government factors in assessing which practices should be applicable.

We synthesized the academic and commercial literature relating to supply chain management practice and, from this synthesis, sought to identify lead-time strategies. Using the results of this literature review as a guide, we then interviewed commercial organizations to learn more about how companies employ these strategies. We chose companies to interview by identifying industry leaders in one or more of the strategies revealed in the literature.

Below, we review the best commercial practices that we identified in the literature. Following this, we discuss our interview methods and the interview findings. Finally, we present recommendations for DLA.

Best Commercial Practices

Current literature on commercial supply chain management practices indicates three key elements of supplier relationship development that affect a company's ability to reduce and control their lead times: negotiating LTCs, managing supplier relationships, and measuring supply chain maturity. Clear strategies to reduce lead times emerged from these themes and could potentially apply to DLA's practices.

Negotiate Long-Term Contracts

A DLA supplier that is not actively managing lead times may simply accept whatever lead time is offered by its suppliers. Alternatively, to actively manage lead times (and thereby minimize them), a supplier can use them as a competitive variable in negotiation with current and prospective suppliers.[1]

The first step in this process is to develop lead-time benchmarks, or a standard lead-time index, for negotiation. The lead-time index provides the expected lead times that can be obtained from suppliers across varying geographical areas and should also vary by commodity type. An example of a lead-time commodity index is shown in Figure E.1.[2]

Through developing this index, a company becomes aware of the gold standard lead times that it can achieve, as well as the average lead time that most suppliers can meet. Having this knowledge puts the company at an advantage in negotiations with future and current suppliers. For current suppliers, the company should communicate that lead times that do not meet the expectations will need to be renegotiated. This can result in eventually changing suppliers or working with that supplier to reach the goal lead time. For future suppliers, negotiations should specify acceptable lead times from the beginning, and final lead-time decisions should be written into the contract.[3]

Figure E.1
Example Lead-Time Commodity Index

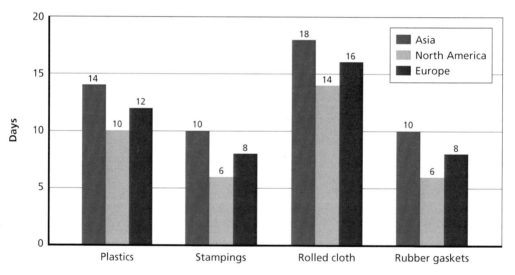

RAND RR822-E.1

[1]   R. Trent and R. Monczka, "Purchasing and Supply Management: Trends and Changes Throughout the 1990s," *International Journal of Purchasing and Materials Management*, Fall 1998; J. H. Perry, "Lead Time Management: Private and Public Sector Practices," *Journal of Purchasing and Materials Management*, September 1990; S. Hafey, "Managing Supplier Relationships is Key to Business Success," *Supply Chain Solutions*, September/October 2010.

[2]   J. Reaume, "6 Procurement Actions That Can Boost Your Business," *Supply Chain Management Review*, 2010.

[3]   Reaume, 2010.

The negotiation should include concessions from both the supplier and the customer, and there should be flexibility in how and when the customer orders and receives the material from the supplier. This allows the customer to effectively respond to the demand it experiences from its customers without being forced to hold higher levels of on-hand inventory. There should also be a maximum period of shipment delay that is fair to the supplier. This allows some flexibility with delivery but still ensures that the customer receives a certain level of support.[4]

Another element of contract negotiation that balances risk between supplier and customer is effectively setting minimum order quantities (MOQs).[5] Higher MOQs allow the supplier to make long production runs in order to realize economies of scale, while lower MOQs enable customers to minimize costs. A lower MOQ allows an organization to make more-flexible adjustments to order the minimum amount of material required to sustain production and respond more closely to the demand it experiences. It also allows the customer to maintain lower inventory levels.[6]

The MOQ also serves as a commitment of a certain level of business with the supplier through the duration of the contract. A higher MOQ therefore can create priority and leverage for a customer with that particular supplier, which allows a customer to negotiate on other points (like lead time) more easily because of the monetary commitment to the supplier. MOQ risks thus need to be balanced between the customer and supplier in order to create and maintain a strong supplier relationship that meets both parties' needs.[7]

Several methods exist to reduce MOQ levels. The first is through contractual adjustments, but this type of adjustment can be very expensive for customers as they attempt to change the original contract agreement. Another option is to authorize the suppliers to make long production runs, but have them ship only the immediate requirements. This method still commits the customers to the same level of orders, but it allows them to spread the delivery over a period of time and maintain lower inventory levels on site, which is valuable if space is at a premium. A third method involves developing minimum order values by consolidating orders for several items with one single supplier. This allows the customer to guarantee the supplier a certain dollar value of business without having to order a set number of each item. This may also be developed across the life of the entire contract, rather than just by order. This allows the customer to still respond in alignment with the demand it experiences, while retaining the same level of commitment to the supplier.[8]

---

[4]   Hafey, 2010; Reaume, 2010.

[5]   Hafey, 2010.

[6]   Reaume, 2010.

[7]   Reaume, 2010.

[8]   Reaume, 2010.

Finally, industry leaders are pushing for broader contracts, either in terms of length or number of items covered. Pratt & Whitney and Lockheed Martin provide two examples.[9] Pratt & Whitney sought to lock in material supplies for the next 30 years by the end of 2012. The contracts they developed locked in suppliers for ten to 20 years rather than the more traditional two- or five-year contracts. The contracts include buffer stocks and guaranteed lead times for the items included on the contract. The purpose of tying in suppliers is primarily to guard against raw material constraints and clearly allows the company to ensure that it will have the supplies it needs within guaranteed lead times.[10] Lockheed Martin has taken the other approach, recently signing a one-year contract with Arrow Electronics for more than 22,000 items. The contract shifted its supply base from more than 240 separate suppliers for these items in 2011 to one single supplier. The contract serves to improve product delivery performance, standardize parts selection, and increase internal efficiency and productivity. More than one-third of the components included in the contract will be delivered with substantial lead-time reductions.[11]

The proper development of LTCs provides the base for developing good supplier relationships. Lead time must be a competitive variable for supplier selection, and minimum order values or quantities must be appropriately set during this phase of supplier relationship development. The current push for industry leaders to develop and strengthen their contract-building practices illustrates the importance of this strategy.

Improve Management of Supplier Relationships

Most private sector strategies for lead-time reduction involve the supplier relationship development process—the long-term management of that relationship after contracts have been negotiated.[12] Through this phase, the company is able to leverage its business with the supplier to collaborate and continuously improve many outcomes, including lead times.

One method to reduce ALT centers on separating the sourcing decision from the buying process.[13] ALT becomes significantly reduced because the work to identify a supplier is completed before the need for the item materializes, and order processing for day-to-day purchasing can be automated. When a PR is made, a supplier is already in

---

[9] Lockheed Martin, "Lockheed Martin Reaches Strategic Supply Chain Agreement," *ENP Newswire*, 2012; R. Miel, "GM 'Metric' System Rates Suppliers," *Plastics News*, Vol. 16, No. 16, 2004; T. Schier, "Engine Maker P&W Looks to Lock in Material Supplies for 30 Years," *Metal Bulletin*, 2012.

[10] Schier, 2012.

[11] Lockheed Martin, 2012.

[12] A. Porter, "Lead Times Are Shrinking, but Not Everyone's a Winner," *Purchasing Magazine*, 1998.

[13] Hafey, 2010; Miel, 2004; P. Teague, "P&G Is King of Collaboration—for Building Close and Productive Relationships Internally and with Suppliers," *Purchasing Magazine*, 2008.

place and the order can be sent directly.[14] A side benefit of separating supplier selection from the buying process is that it can also facilitate supply base rationalization efforts.[15]

Appropriately defining the buying method for different types of products can also help reduce ALT. When a limited number of vendors for certain items are available, the companies should establish LTCs. This ensures access to required materials and enables using monetary commitments to negotiate for better lead times, as noted above. For products with high and stable demand, companies should implement multiyear buying strategies through competitive sourcing. For these types of products, actual supplier performance (discussed below) can be used to allocate business.[16]

Many of the items procured by DLA have low and unstable demand. Information from the commercial sector about buying strategies for these types of items is scarce. A previous RAND study conducted for the Air Force identified supply strategies for low-demand service parts and found that the best time for developing supply strategies for low-demand parts is before acquisition of the supported end item begins. This is when the buyer has the most leverage with its suppliers and can align the incentives to motivate the supplier to provide long-term support.[17]

Regardless of the type of item, the process of buying should be automated and standardized to the extent that the market allows.[18] This of course varies based on the supplier market. Some suppliers may not have the electronic systems in place that allow for automated buying practices. Automating the ordering process minimizes disputes over pricing and delivery dates and quickens order transmission by streamlining the purchasing process.[19]

Collaborating on initiatives that benefit both the supplier and the customer are an important part of supplier relationship management. Collaboration requires both the customer and the supplier to be open to suggestions offered by the other, especially in terms of improvement to their processes. Industry leaders have demonstrated that by becoming a better customer, organizations can find that suppliers are more willing to work with them to meet lead times or other needs during negotiations.[20] One example of a collaboration initiative is developing electronic systems that facilitate information-

---

[14] Perry, 1990.

[15] Trent and Monczka, 1998; Hafey, 2010; S. Avery, "Purchasing Salutes Suppliers; World-Class Companies Use Recognition Programs as a Tool to Develop and Reward Preferred Suppliers," *Purchasing Magazine*, 2008; W. Morris, "How to Leverage Supplier Performance Management for Continuous Supply Chain Improvement," *Supply & Demand Chain Executive*, May/June 2010.

[16] Perry, 1990.

[17] M. Chenoweth, J. Arkes, and N. Moore, *Best Practices in Developing Proactive Supply Strategies for Air Force Low-Demand Service Parts*, Santa Monica, Calif.: RAND Corporation, MG-858-AF, 2010.

[18] Trent and Monczka, 1998; Perry, 1990.

[19] Hafey, 2010.

[20] *Purchasing Magazine* Staff, "Spotlight Shines on Key Suppliers," *Purchasing Magazine*, 2009.

sharing and order automation.[21] Collaboration can focus on other areas as well and can lead to continuous improvement of lead time, cost, quality, and other aspects of the supplier relationship.[22]

This kind of collaboration has shown strong effects on reducing lead times. In one study, customers that employed supplier relationship management techniques had PLTs that averaged six days, compared with 20 days for those who did not employ such techniques.[23] The reduction in PLT results from better forecasting, which is possible with information-sharing. Lead-time reductions also result from the opportunities to improve production and procurement processes that are created in this environment. Companies managing their suppliers in this way take more-direct action to help develop supplier capability and performance.[24] Potential process improvements focus on improving the management of the production line by reducing the inventory in process, updating to better machinery that reduces production time, and cross-training employees to ensure that staff availability is not the cause of bottlenecks.[25] Another area for improvement ties back to contract negotiation: A customer can offer to pay more for a shorter lead time. The lead time can also be set to allow the supplier to have enough flexibility to fit in potential rush orders.[26]

Measurement is another key aspect of supplier relationship management.[27] Industry leaders use supplier scorecards to monitor supplier performance on metrics such as lead time, on-time delivery, quality, and cost.[28] The qualitative measures allow suppliers to provide an explanation for why processes fall out of control, such as reasons for late delivery.[29] Organizations should clearly communicate these key metrics with their suppliers so that the suppliers have an understanding of what is important to the client.

---

[21] R. Spiegel, "SRM Leaders Outpace Peers On Lead Times, Other Key Metrics," *Supply Chain Management Review*, January/February, 2011; Hafey, 2010.

[22] Perry, 1990; Hafey, 2010.

[23] Spiegel, 2011.

[24] Trent and Monczka, 1998; Teague, 2008.

[25] Porter, 1998; W. Hopp, M. Spearman, and D. Woodruff, "Practical Strategies for Lead Time Reduction," *Manufacturing Review*, Vol. 3, No. 2, 1990; K. J. Youngman, "A Guide to Implementing the Theory of Constraints," web page, 2009; S. Hiiragi, *The Significance of Shortening Lead Time from a Business Perspective*, Discussion Paper Series, No. 391, Manufacturing Management Research Center, University of Tokyo, March 2012; M. Wouters, "Economic Evaluation of Lead Time Reduction," *International Journal of Production Economics*, 1991; J. Olson, "Lead Times: It's All About the Process," *Surface Fabrication*, September 2009; "Making the Connections," *Metalworking Production*, Vol. 153, No. 2, March 2009; M. Collins, "QRM for Reducing Lead Times," *Industrial Maintenance and Plant Operation*, December 2008; "Palletized Production Reduces Lead Times," *Quality*, March 2011.

[26] Youngman, 2009.

[27] Perry, 1990; Hafey, 2010; Spiegel, 2011.

[28] Hafey, 2010; Spiegel, 2011.

[29] Morris, 2010.

Supplier performance should also be communicated formally with the understanding that underperformers will work to address limitations or the company may consider replacing them.[30] By using these metrics, a company can ensure that suppliers continue to meet the lead times required in their contracts, but it can also highlight areas for improvement that may serve to reduce lead time.

In addition to communicating supplier performance metrics, organizations should share other information with suppliers. This includes demand and inventory information that allows suppliers to better plan and respond to the demands of the company, including being better able to deliver the items within the lead time.[31]

In addition to scorecards, industry leaders recognize outstanding performance to motivate other suppliers to exceed expectations. Table E.1 shows a summary of leading companies and the metrics they use for their supplier performance awards. These include quality, on-time delivery, cost, responsiveness (i.e., the times of different production elements), overall performance, and lead time, though not all companies use

Table E.1
Metrics Used for Supplier Performance Awards

Metrics	Quality	On-Time Delivery	Cost	Responsiveness	Overall Performance	Other	Lead Time
Procter & Gamble	X	X	X	X	X	X	X
Analog Devices	X	X	X	X	X		
Intel	X	X	X	X			
Rockwell Collins	X	X	X			X	
Cessna	X	X	X			X	
Texas Instruments	X		X	X	X	X	
United Technologies Corporation	X	X		X			
General Motors	X	X				X	X
Northrop Grumman	X	X				X	
BF Goodrich	X	X				X	
Emulex	X	X					X

SOURCES: Miel, 2004; Teague, 2008; Avery, 2008; *Purchasing Magazine* Staff, 2009; Day, 2008; Goodrich Corporation, 2009; Flextronics, 2012; KEMET Corporation, 2012.

NOTE: The marked cells denote metrics for which the corresponding company gives awards for supplier performance.

---

[30] Hafey, 2010; Morris, 2010; Spiegel, 2011; R. Trent, "Creating the Ideal Supplier Scorecard," *Supply Chain Management Review*, March/April 2012.

[31] Perry, 1990; Hafey, 2010; Trent, 2012.

all metrics.[32] Some also include metrics for certain product types, such as environmental regulations. All of these industry leaders award suppliers for providing a quality product, and all but one of these leaders highlight the importance of on-time delivery, which is a clear substitute for lead-time performance. By ensuring that the product is delivered on time, the company ensures that at the very least, the negotiated lead time has been met. However, including absolute lead time as an element of supplier performance awards is on the cutting edge, with only a handful of firms distinguishing it from the more general measure of on-time delivery.

Measure Supply Chain Maturity

The final aspect of supplier relationship development that is related to lead time is an organization's level of supply chain maturity. By measuring this, an organization can identify areas for lead-time reductions. Supply chain maturity ranges from providing the basic functions to support a supply chain process to full cross-collaboration and integration with supplier partners. Maturity is primarily defined by the level of information exchange and the expertise in the wide range of supply chain tasks. It is usually defined in stages, and an organization must meet all requirements of the previous stage across all parts of the supply chain to advance to a more developed stage. Full maturity implies that processes are well understood across the supply chain, supported by documentation and training, consistently applied, and continuously improved. By assessing its current maturity stage, an organization gains an understanding of current practices and can identify areas for improvement, which often include improvements that would significantly reduce lead time.[33] Various supply chain maturity models differ in their exact definitions of each maturity stage, but they follow a similar pattern.[34] Figure E.2 provides one example of supply chain maturity stages.

In addition to stages of maturity, some models include opportunities for improvement by defining key performance metrics. For example, the Supply Chain Operations Reference Model defines five key attributes: reliability, responsiveness, agility, costs, and assets (see Table E.2). Reliability and responsiveness are the attributes most closely related to lead-time improvement. Reliability refers to how closely a supplier meets the expected lead time, and responsiveness measures the cycle times that make up the varying components of lead time.

---

[32] Miel, 2004; Teague, 2008; Avery, 2008; *Purchasing Magazine* Staff, 2009; J. Day, "Suppliers Take Center Stage," *Purchasing Magazine*, 2008; Goodrich Corporation, "Goodrich Presents US Tool Group with Strategic Supplier Award," *Business Wire*, 2009; Flextronics, "Flextronics Receives Emulex Supplier of the Year Award," *PR Newswire*, March 2012; KEMET Corporation, "KEMET Receives Rockwell Collins 2012 Top Supplier Award," *PR Newswire*, 2012.

[33] M. Lahti, A. Shamsuzzoha, and P. Helo, "Developing a Maturity Model for Supply Chain Management," *International Journal of Logistics Systems and Management*, Vol. 5, No. 6, 2009.

[34] Lahti, Shamsuzzoha, and Helo, 2009.

Figure E.2
Supply Chain Maturity Stages Model

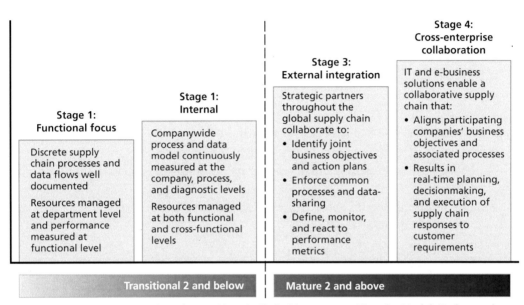

SOURCE: PRTM Management Consultants, "Supply Chain Management Maturity Model: Understand the Transformation Required to Move from a Functionally Focused Supply Chain to Cross-Enterprise Collaboration," 2005.
RAND RR822-E.2

Table E.2
Key Measures for Varying Supply Chain Attributes

Attribute	Key Measures	
Reliability	• Perfect order fulfillment • Percent orders delivered in full • Delivery performance to customer commit date	• Documentation accuracy • Perfect condition
Responsiveness	• Order fulfillment cycle time • Delivery cycle time • Source cycle time (internal and external)	• Make cycle time • Delivery retail cycle time
Agility	• Upside and downside supply chain flexibility • Upside and downside supply chain adaptability	
Costs	• Supply chain management costs • Costs to plan, source, make, deliver, return • Indirect cost related to production	• Cost of goods sold • Mitigation cost • Direct labor/material cost
Assets	• Cash-to-cash cycle time • Inventory days of supply • Return on supply chain fixed assets	• Days sales outstanding • Days payable outstanding • Return on working capital

SOURCE: Supply Chain Council, *Supply Chain Operations Reference (SCOR) Model*, Version 10, 2010.

Best Practices for Reducing Lead Times

We have identified specific practices associated with reducing lead times, which we summarize below.

- Negotiate LTCs.
  - Factor lead times into selection.
  - Include lead times in negotiations and final contract terms.
  - Balance minimum order quantities to share risk between the supplier and customer.
- Improve management of supplier relationships.
  - Separate the sourcing decision from the buying process.
  - Include lead time as an element of supplier scorecards, which should be used for supplier selection.
  - Share information and automate processes.
  - Work with suppliers to reduce PLT.
- Measure supply chain maturity.
  - Identify areas for improvement.
  - Focus efforts on appropriate performance measures.

Interview Methodology

Based on our synthesis of the literature, we developed a structured protocol to interview industry leaders in supply chain management. The protocol included eight primary questions regarding sourcing strategies, supplier relationship management, PLT, and ALT (Table E.3). We developed prompts for each question to facilitate rich responses from the interviewees. The questions were designed to understand how companies actually employ the best practices identified in the literature. The answers provide an indication of whether the strategies are effective in practice or not.

Table E.3
Interview Protocol

Focus Area	Primary Questions Asked
Sourcing strategies	• When your company is in the process of selecting suppliers for a new contract, does supplier lead time factor into that selection?
Supplier relationship management	• Do you measure the performance of your suppliers in terms of lead time? Please describe. • Is lead time a part of contract negotiations? • Does a process exist in which the suppliers can rate your performance as a customer?
Production lead time	• Do you work with suppliers to improve their lead-time performance? • What type of production lead time do you generally experience?
Administrative lead time	• What types of strategies do you employ to reduce administrative lead time? • What type of administrative lead time do you generally experience?

The interviews targeted leading commercial firms that employ these best practices and operate in industries that overlap with DLA's product sectors. Table E.4 shows these industries and the number of leading firms in each. The companies identified were cross-referenced with DLA's top suppliers. For firms that did business with DLA, that business is conducted by a government business division, which is separate from the firms' commercial divisions. The interviews were all conducted with the commercial side of these firms.

The interviews were conducted over the phone and took approximately 30 minutes to complete. At least two researchers participated in each call to allow for transcription-style notetaking to take place during the interview. This allowed us to avoid interviewer bias in the results by analyzing direct quotes rather than interpretations of the discussion. Responses were analyzed not only across a single company but also across companies for question groups to more easily identify recurring themes or exceptions to the norm in each focus area. Final analysis of the interview findings employed a pile sorting method. We recorded each observation on a piece of paper, and then arranged these into related piles. We then labeled each pile based on the groupings to identify clear themes from the responses. This method of analysis allowed themes to emerge organically and ensured that observations were not overlooked in the final results. Finally, we vetted themes across multiple members of the research team to minimize unintended researcher bias.

Interview Findings

We contacted 30 companies via email, with multiple follow-up contacts to increase participation. There were eight completed responses to the primary set of questions—five telephone interviews and three written responses. Twelve firms did not respond in any manner, and seven firms declined to participate outright. For the remaining firms, we were unable to reach a person able to answer our questions. After the first four interviews, we added a line of questioning relating to MOQs to the protocol because our interest in the ability of commercial companies to balance and manipulate this value increased as we saw the responses to the initial interviews.

Table E.4
Number of Leading Firms in Industries Comparable to DLA

Industry	Number of Firms
Aerospace and aviation	9
Automobile and car parts manufacturing	3
Equipment manufacturing	2
Provision of consumer goods to market	7
Semiconductor and computer technology	9

The companies we interviewed were relatively similar in their reports of which practices they implement more and which they implement less. The strategies that were consistently implemented may be easier to implement or have lower implementation costs than the strategies implemented less consistently. Full results of the interview analysis are presented in Table E.5, with each practice rated as low, moderate, or consistent based on its level of implementation among the companies interviewed.

Consistently implemented best practices include factoring lead time into supplier selection, separating the sourcing decision from the buying process, sharing information and automating processes, and focusing efforts on the appropriate performance measures. Lead time was a primary driver for supplier selection for all of the companies interviewed. The companies were aware of what types of lead times they can achieve from a variety of suppliers and in turn factor this into negotiations. By separating the sourcing decision from the buying process, companies were able to reduce their ALT because the ordering process was automated. Information was easily shared with suppliers, and processes could be automated because the sourcing decision was already made. Companies also reported that they consistently evaluated performance measures to ensure that change efforts were focused appropriately. Many reported shifting their performance evaluations to focus more highly on speed (and therefore lead time) in an effort to remain competitive.

There was moderate evidence for including lead time in contract terms. Although lead time was a primary factor in supplier selection, the exact terms were not always explicitly written into contracts. Companies preferred to maintain an open line of

Table E.5
Best Practice Implementation Among Interviewed Companies

Best Practice	Use
Negotiate LTCs.	
Factor lead time into selection.	Consistent
Include lead time in contract terms.	Moderate
Balance minimum order quantities.	Moderate
Improve management of supplier relationships.	
Separate sourcing decision from buying process.	Consistent
Include lead time as part of supplier scorecards.	Low
Share information and automate processes.	Consistent
Work with suppliers to reduce PLT.	Moderate
Measure supply chain maturity.	
Identify areas for improvement.	Moderate
Focus efforts on appropriate performance measures.	Consistent

communication so that if a supplier was going to be late on an order, it felt comfortable notifying the company, which allows the company time to make other arrangements for that order or plan accordingly for its delay. This was often as successful as forcing penalties on a supplier for late deliveries because it developed a relationship of trust. For the practice of balancing MOQs, it is typically suppliers, rather than the customer companies, that bring this up during negotiations. The use of balancing strategies depended on the type of item. The remaining two moderately employed strategies included working with suppliers to reduce PLT and identifying areas for improvement. We ranked these strategies as having a moderate implementation level because companies tend to address these practices only when a problem arises.

Although somewhat common in the literature, our interview responses did not indicate much use of lead time as an explicit part of supplier scorecards. Some companies were aware that lead time would be a good measure but had not determined an accurate way to track it. Two companies did include lead time directly in their supplier scorecards; the remaining companies factored in related measures, such as on-time delivery.

In addition to questioning the companies about best practices employed, we also asked about ALT and PLT estimates. For those companies that were willing to respond, the typical ALT was one day, with a maximum of three days for an item that is purchased more rarely. The shortened ALT was attributed to completing the supplier selection process at the front end, separate from the ordering process. This made it possible to automate the day-to-day ordering needs for production managers. PLT estimates had quite a bit of range, from between two and five days to between seven and eight weeks, depending on the item type. No PLT estimates higher than eight weeks were reported. Although not directly, supply base rationalization was reported as contributing to the low PLT estimates. By having a reduced set of suppliers, companies can more easily negotiate shorter lead times. Separating supplier selection from ordering yields a timeline that has very short ALT compared with PLT, as shown in Figure E.3.

Figure E.3
Lead-Time Timeline

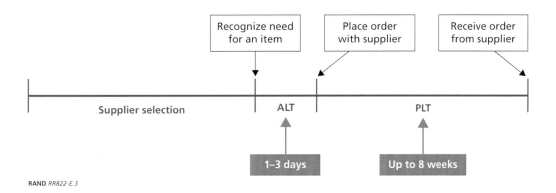

Recommendations for DLA

Combined with our examination of supply chain agility in DLA, the findings here suggest that many commercial best practices for reducing lead times can be used by DLA. Some of these best practices may be easier for DLA to implement quickly because commercial organizations appear to implement these practices consistently. Consistently used practices may be explained because they are easiest to implement, or the cost-benefit of implementation may provide the justification. These strategies include the following:

- Factor lead time into supplier selection.
- Separate sourcing decision from buying process.
- Share information and automate processes.
- Focus efforts on appropriate performance measures.

Best practices used less frequently may be later steps in the effort to reduce lead times. These include the following:

- Include lead time in contract terms.
- Work with suppliers to reduce PLT.
- Identify areas for supplier improvement.
- Include lead time as part of supplier scorecards.

Department of Defense and Defense Logistics Agency Policy

After the initial look at commercial best practices, we did a similar survey of existing DoD and DLA policy. DoD and DLA publish policy documents that address a wide range of required approaches and procedures to which stakeholders must adhere for supply chain material management. This guidance is designed to standardize practices around those determined to be most effective and to ensure that statutory requirements are met.

In this appendix, we

- integrate the categories of commercial best practices with the review of both DoD and DLA policy, focusing on those sections relevant to lead-time management
- use the commercial best practices as a guide for what aspects of policy to review and what features to look for
- identify gaps in policy where additional guidance may help reduce lead times
- examine both upstream and downstream supply chain management policy.

Note that both DoD and DLA policies address many best practices in purchasing and supply chain management, and this review is not an assessment of most of these. Rather, this review focuses only on best practices in reducing lead times.

Upstream Best Practices

Upstream best practices are those that refer to DLA's interaction with its suppliers, or the acquisition components of DLA. We categorize upstream best practices for analyzing DoD and DLA policy into three groups, based on the themes found in the commercial literature and described in Appendix E:

- *Negotiate LTCs*: Factor lead time into supplier and bid selection, include lead time in contract terms, and balance customer and supplier risk in establishing minimum order quantities.

- *Improve management of supplier relationships*: Separate the sourcing decision from the buying process, include lead time as part of supplier scorecards, share information and automate processes, and work with suppliers to reduce PLT.
- *Measure supply chain maturity*: Identify areas for process improvement and focus improvement efforts on appropriate performance measures.

The review of DoD and DLA guidance revealed that few policies address these upstream best practices directly, though several policies address related issues. LTCs and supplier relationship management are both recurring topics in policy, though not necessarily with a focus on lead times. Policy does not explicitly address analyzing internal supply chain maturity, but we note that both DoD and DLA maintain a culture of continuous process improvement and have ongoing efforts to improve the supply chain.

DoD and FAR Policy and Upstream Supply Chain Best Practices

We start by examining DoD policy and comparing it with best practices. We reviewed DoD 4140.1-R, *DoD Supply Chain Materiel Management Regulation*;[1] DoD Directive 5000.01, *The Defense Acquisition System*;[2] and selected portions of the FAR.[3] In general ways, the regulations support tracking metrics on suppliers, limiting lead times, managing order quantities, considering holding costs, and coordinating with customers. The documents present more-specific guidelines for ensuring that supply levels match demand, for minimizing logistics footprint, and for making sourcing decisions on best value. Specific language found in support of each best practice is summarized at the end of this appendix. We classified how comprehensively the policy addresses each area, and the results are summarized in Table F.1.

Table F.1 uses symbols to indicate the completeness of DoD policy for each of the identified best practices. A blank circle indicates that our team was unable to identify any DoD-level policy that clearly addresses the best practice. Partial policy is indicated with a bolded circle.

DLA Policy and Upstream Supply Chain Management Best Practices

After reviewing DoD policy, we looked at DLA policy, including the Defense Logistics Acquisition Directive (DLAD) *SMSG Business Rules* and the initial briefing for the Superior Supplier Incentive Program.[4] We also scanned several DLA issuances that

---

[1] U.S. Department of Defense, *DoD Supply Chain Materiel Management Regulation*, Office of the Deputy Under Secretary of Defense for Logistics and Materiel Readiness, DoD 4140.1-R, May 23, 2003b.

[2] U.S. Department of Defense, *The Defense Acquisition System*, Directive 5000.01, May 12, 2003a.

[3] Federal Acquisition Regulation, 48 C.F.R., 2014.

[4] Defense Logistics Agency, *SMSG Business Rules*, Revision 5, Defense Logistics Acquisition Directive, September 9, 2013; DLA, "DLA Strategic Sourcing Briefing: DoD Superior Supplier Incentive Program (SSIP)

Table F.1
Comprehensiveness of Upstream Best Practices in DoD and DLA Policy

Best Practice	Comprehensiveness of Policy	
	DoD	DLA
Negotiate long-term contracts.		
Factor lead time into selection.	○	●
Include lead time in contract terms.	○	●
Balance minimum order quantities.	●	○
Improve management of supplier relationships.		
Separate sourcing decision from buying process.	●	●
Include lead time as part of supplier scorecards.	○	●
Share information and automate processes.	●	○
Work with suppliers to reduce PLT.	○	●
Measure supply chain maturity.		
Identify areas for improvement.	●	○
Focus efforts on appropriate performance measures.	●	○

NOTE: A blank circle indicates that our team was unable to identify any policy that clearly addresses the best practice. Partial policy is indicated with a bolded circle.

provide further information on the guiding principles for employees. Note that DLA policy does not necessarily need to address each best practice; it only needs to supplement DoD policy for DLA's contracting staff to have sufficient guidance.

The review of DLA policy and training documentation found that DLA does address some best practices related to lead times, though often in an indirect way. Greater emphasis is placed on collaboration between acquisition stakeholders (such as contracting officers and strategic supplier groups) and their suppliers than on lead times. The results of this policy review are summarized in Table F.1 alongside the results of the DoD-level policy review.

The review indicates that best practices identified in the commercial literature all have some coverage in either DoD or DLA policy. None of this coverage is complete, but there is precedent or foundation to incorporate lead times into processes in each of the ways identified above.

Initial Meeting," internal briefing by G. L. Starks, February 18, 2011, provided to RAND by DLA.

Areas for Policy Expansion

After reviewing existing DoD and DLA policy, we found two potential areas for policy development involving upstream best practices related to reducing lead times: incorporating consideration of PLT into existing supplier processes and incorporating PLT reduction efforts into internal DLA assessments.

Policies addressing all supplier-related processes—including supplier selection, contract development, and supplier management—need an increased emphasis on lead times. This often means moving policy from general references to using PLT as an explicit criterion where possible. Using PLT as part of supplier selection and supplier relationship management is the place to start. Because of the possibility of protests from bidders who did not win a contract, including PLT as a contractual term may need to be focused only on certain contracts of sufficient value to make this burden worthwhile with a positive return on investment.

DLA could also benefit from tracking its efforts to reduce PLTs. Policy that specifies analysis of PLTs over time, especially within SSAs, can help identify which efforts are most effective.

Specific Policy References: Upstream Best Practices

We have selected some of the most pertinent passages identified in DoD and DLA policy that relate to the supply chain management best practices associated with lead-time management.

DoD and FAR Policy
Negotiate Contracts

- Factor lead time into selection.
  - "The DoD Components shall aggressively pursue the lowest possible acquisition lead times."[5] While most of the lead-time policy seems to accept supplier lead times as a given and guide DoD components to plan for this in their acquisition process, this phrase is an exception.
  - "[I]n determining cost-effectiveness of stockage alternatives, the DoD Components shall include all applicable elements of cost and cost savings (e.g., inventory holding costs, and second destination transportation) in determining responsiveness, including timeliness."[6]

---

[5]   DoD, 2003b, C2.6.3.1.4.2.

[6]   DoD, 2003b, C3.2.3.5.

- "Analytical and audit support tools shall be developed to aid in considering quantity and/or price and lead time data with other relevant data so that contract award decisions are based on the best value to the Government."[7]
- "Satisfy the customer in terms of cost, quality, and timeliness of the delivered product or service."[8]
- Policies addressing use of past performance for selection provide a foundation for emphasizing lead time as a criterion. Examples of these foundational polices include:
  - Consider past performance information prior to making a contract award.[9]
  - "Consider all costs associated with materiel management . . . in making best value logistics materiel and service provider decisions."[10]
  - "The DoD components shall establish criteria and methods to identify contractors who consistently fail to meet contract requirements and prevent future contract awards to such contractors."[11]
- Include lead time in contract terms.
  - No policies identified.
- Balance minimum order quantities.
  - "When EOQ methods are used, every attempt shall be made to purchase materiel under indefinite delivery and indefinite quantity (IDIQ) contracts, so the order quantity and delivery times are reduced."[12]

Improve Management of Supplier Relationships
- Separate sourcing decision from buying process.
  - "When EOQ methods are used, every attempt shall be made to purchase materiel under indefinite delivery and indefinite quantity (IDIQ) contracts, so the order quantity and delivery times are reduced."[13]
- Include lead time as part of supplier scorecards.
  - Use contractors who have a track record of successful past performance or who demonstrate a current superior ability to perform.[14]
- Share information and automate processes.

---

[7] DoD, 2003b, C4.3.2.4.

[8] FAR, 2014, 1.102 (b) (1).

[9] FAR, 2014, 9, 13, 15, 36, and 42.

[10] DoD, 2003b, C1.3.1.3.

[11] DoD, 2003b, C3.5.2.7.

[12] DoD, 2003b, C2.6.3.1.2.

[13] DoD, 2003b, C2.6.3.1.2.

[14] FAR, 2014, 1.102 (b) (1) (ii).

- "The DoD Components may extend collaborative forecasting to commercial suppliers to improve the support that those suppliers provide."[15]
- Contractors can see their past performance ratings in PPIRS, and no new systems should be built for this purpose.[16]
- "Provide for visibility of the quantity, condition, and location of in-storage, in-process, and in-transit assets throughout the DoD supply chain visibility of orders placed on organic and commercial sources of supply."[17]
- "Integrate information exchange between materiel managers and sourcing and acquisition managers."[18]
- "Communication with customers, including performance feedback, should occur throughout sourcing and acquisition processes."[19]
- Work with suppliers to reduce lead times.
  - Use performance-based acquisitions, logistics strategies, and agreements.[20]
  - Develop and maintain metrics that address enterprise functional and process levels of operations.

Measure Supply Chain Maturity
- Identify areas for improvement.
  - "DoD components shall use metrics to evaluate the performance and cost of their supply chain operations."[21]
- Focus areas on appropriate performance measures.
  - "DoD components shall use metrics to evaluate the performance and cost of their supply chain operations."[22]

DLA Policy
Negotiate Contracts
- Factor lead time into selection.
  - As part of acquisition plans, the DLAD *Procedures, Guidance, and Information* says to "provide estimates of production lead times" and, within competition,

---

[15] DoD, 2003b, C2.5.1.5.

[16] PPIRS data fields provided to RAND by DLA.

[17] DoD, 2003b, C1.3.12.

[18] DoD, 2003b, C3.1.1.2.3.

[19] DoD, 2003b, C3.1.2.1.1.

[20] DoD, 2003a, E.1.1.16; DoD, 2003b, C1.3.2.1.

[21] DoD, 2003b, C1.5.1.

[22] DoD, 2003b, C1.5.1.

"discuss the trade-offs of use of such substitutes in terms of price differences, quality, and acquisition and production lead time."[23]

- Policies addressing use of past performance for selection provide a foundation for emphasizing lead time as a criterion. Examples of these foundational polices include:
  - "The Defense Logistics Agency (DLA) will evaluate a Contractor's past performance, including, but not limited to, their record of conforming to specifications, conformance to the standards of good workmanship, adherence to contract schedules, and commitment to customer satisfaction. DLA utilizes the following information systems in evaluation of Contractor past performance: automated best value system (ABVS); past performance information retrieval system—statistical reporting (PPIRS-SR)."[24]
  - Strategic Materiel Sourcing Groups monitor supplier performance metrics on strategic-level contracts.[25]
  - "When used in competitive negotiated best value source selections, past performance information will be evaluated based upon each offeror's demonstrated recent and relevant (relevancy) record of performance in order to reach the final performance confidence assessment rating. Contracting officers are advised not to rely solely on the ABVS, PPIRS-Statistical Reporting, PPIRS-Report Card or other performance assessments/ratings, and should consider reviewing the data used to construct the performance score if the circumstances of the procurement dictate."[26]

- Include lead time in contract terms.
  - No policy references found.
- Balance minimum order quantities.
  - Order quantities are mentioned, but not with regard to balancing them as a trade-off with other contract elements (e.g., price).

Improve Management of Supplier Relationships

- Separate sourcing decision from buying process.
  - Shift to LTCs where possible and evaluate LTCs with LTC Project Tracker.[27]
- Include lead time as part of supplier scorecards.
  - DLA policy does not prevent including this, but it does not explicitly mention PLT as a performance characteristic: "Contracting officers are not precluded

---

[23] Defense Logistics Agency, *Procedures, Guidance, and Information (PGI)*, Defense Logistics Acquisition Directive, December 4, 2012, section 7.105-90.

[24] DLA, 2013, 52.213-9005 (a) (1), p. 624.

[25] DLA internal briefing, 2011.

[26] DLA, 2012, 15.304-90 (b).

[27] DLA, 2012, 16.190.

from collecting/analyzing past performance information in addition to ABVS/ PPIRS-SR."[28]
- Share information and automate processes.
  - There is substantial policy guidance on automated solicitation and awards, but not on automated processes with suppliers that result from sharing information.
- Work with suppliers to reduce lead times.
  - "Performing analysis and resolving post award requests from the supply planner . . . or customer account specialist . . . ensures correct products are delivered and achieve[s] current timeline, accuracy, customer satisfaction and administrative and production lead time (ALT/PLT) targets."[29]

Measure Supply Chain Maturity
- Identify areas for DLA improvement.
  - No policy references found.
- Focus areas on appropriate performance measures.
  - No policy references found.

---

[28] DLA, 2013, 13.106-2 3iiDID, p. 190.

[29] DLA, 2013, 42.1103-90.

Lead-Time Analysis

A key factor in improving DLA supply chain agility is reducing lead times. The lead time for an order consists of two components. The first is *administrative lead time* (ALT), which is time from the requisition date to the award date; the second is *production lead time* (PLT), which is the time from the award date until the delivery date. The *delivery date* is defined for DLA measurement as the date when the percentage of the order received exceeds 50. We analyzed the data in an effort to identify factors that drive differences in lead times, including method of procurement, NIIN characteristics, and frequency of procurement. These factors may help to explain differences in lead times among DLA supply chains. While such a statistical analysis cannot explain the reasons for the observed differences, it can point to areas for further investigation by those conducting lead-time reduction efforts, such as the DLA Time to Award team. And it can be useful in understanding metrics used to drive continuous improvement, enabling segmentation of the metrics; this is an area where there cannot be common standards or goals, given differences in supply chain characteristics.

Data Processing Details

The data used in this analysis come from the DLA Active Contracts file, BPUR2 Purchase Order Line table. Extracts from these data were provided to RAND by DLA Headquarters with the assistance of DORRA. The data consist of contract lines from contracts with award dates between September 1, 2010, and August 31, 2012. Because the purpose of this study was to improve DLA supply chain agility for items held in inventory, we excluded from our analysis records that were identified as being direct vendor delivery. We further excluded records that appeared to have missing or incorrect data fields. For the ALT/PLT analysis, we excluded from our analysis contract lines where the award date was earlier than the requisition date (which would result in a negative ALT), or where the delivery date was earlier than the award date (which would result in a negative PLT). In addition, we only retained records in which the purchase method—manual versus automated—could be positively identified by DORRA; records with missing values were excluded from our analysis. The result was approximately 1.22 million contract lines with ALT information and 1.11 million with PLT information. We used these data points for the majority of analysis, except when otherwise described below.

Manual Versus Automated and LTC Versus Non-LTC Purchases

For virtually all of the statistical analysis that was performed, we separated the contract line data into one of the following three groups of order method: purchases of items that are on LTCs that are made via automated systems, which we have termed *automated LTC*; purchases of items that are not on LTCs but that are made via automated systems, which we have termed *automated non-LTC*; and purchases that are *manual*. As Figures G.1 and G.2 show, the median ALT and PLT among automated orders is far lower than for manual orders, and there clearly exists a further difference between automated LTCs and automated non-LTCs, with the former having a lower median ALT, albeit higher PLT. Therefore, we considered these three groups separately for analysis of other variables, based on their median lead times (although the differences in median PLT were less pronounced than for median ALT). This decision helped prevent conflating the effect of automated/manual or LTC/non-LTC with the effect of other variables of interest on ALT and PLT. Additionally, from a bias/variance trade-off perspective,[1] the fact that each of the three groups had relatively similar sample sizes was ideal.

After accounting for automated versus manual and LTC versus non-LTC, we analyzed ALTs and PLTs according to their differences by supply chain, federal supply group (FSG), NIIN recently awarded or not, unit cost, order quantity, and contract dollar range. Although all of these variables were found to have an effect, different variables had more or less effect than others, as this section of the appendix details.

Analysis by Supply Chain

The first variable analyzed (after accounting for the automated versus manual and LTC versus non-LTC factors) was supply chain, first by ALT and then by PLT. To describe the population being analyzed, there were five major supply chains (those with significant numbers of items held in inventory): Aviation, Construction and Equipment,

---

[1]   The bias/variance trade-off is an ever-present aspect of regression analysis concerning the concept of overfitting and underfitting. On one hand, adding more variables will almost always lower the bias involved with a regression estimate. However, adding more variables will typically increase the variance associated with any prediction that can be formed from the model. When data sets have dramatically unequal sample sizes for, say, three major groups (e.g., 80/15/5 percentage splits), this creates challenges from a model building perspective because the sample sizes for certain strata can become too small to draw meaningful inference when further subcategorized.

Figure G.1
Number of Contract Lines and Median Administrative Lead Time, by Order Method

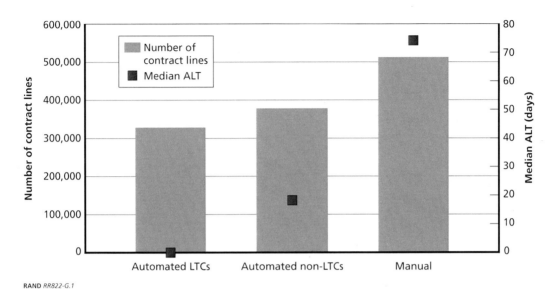

Figure G.2
Number of Contract Lines and Median Production Lead Time, by Order Method

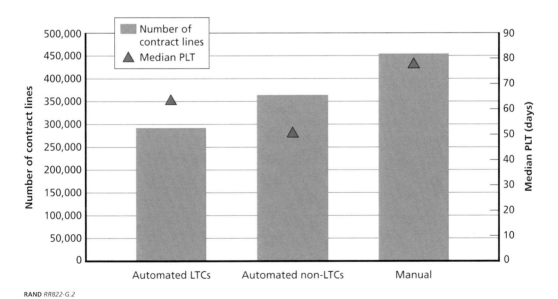

Industrial Hardware, Land, and Maritime. Each had at least 10,000 contract lines that fell into each of the three groups: automated LTC, automated non-LTC, and manual. Figure G.3 shows the number of contract lines. Although Construction and Equipment, Industrial Hardware, and Land all have roughly similar percentages of lines

Figure G.3
Number of Contract Lines, by Supply Chain and Order Method

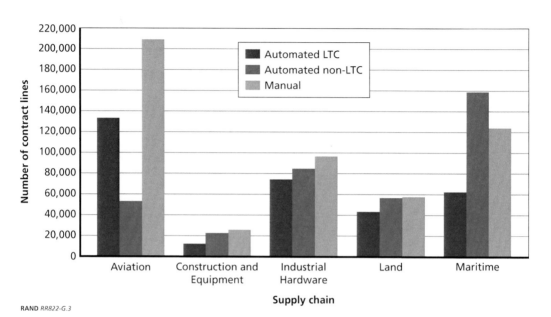

for the three major contract category groups, Aviation clearly has a higher percentage of manual orders, and Maritime has the largest percentage of automated non-LTCs. These differences in types of contracts used underscore the importance of separating the three major groups before drawing conclusions about supply chain behavior, as inherent differences undoubtedly occur between automated and manual orders.

Given the large sample size for all 15 groups indicated in Figure G.3, each of the three major groups could be studied in isolation by supply chain. Figures G.4, G.5, and G.6 show ALT percentiles by supply chain for automated LTCs, automated non-LTCs, and manual lines, respectively.[2]

For automated LTCs and automated non-LTCs, Aviation's ALTs were far longer than other supply chains on average, but Figures G.4 and G.5 explain more about the effect of Aviation's upper percentiles on their mean ALTs. (Note that Figures G.4, G.5, and G.6 are on different scales, due to the substantial difference in lead times for the different ordering methods.) All five supply chains show almost exactly the same 25th and 50th percentiles, but differ more at the 75th percentile; for instance, Aviation's 75th percentile in both Figure G.4 and G.5 is substantially longer in days. Although Figures G.4 and G.5 do not indicate percentiles higher than the 75th, further analysis found that the difference in ALT among Aviation and other supply chains becomes

---

[2] To alleviate the distortion of averages by outliers in the right tail of the distributions, we applied right-censors to the ALT data. We censored automated LTC orders at 20 days, automated non-LTC orders at 50 days, and manual orders at 600 days.

Figure G.4
Administrative Lead Time Statistics for Automated LTC Orders, by Supply Chain

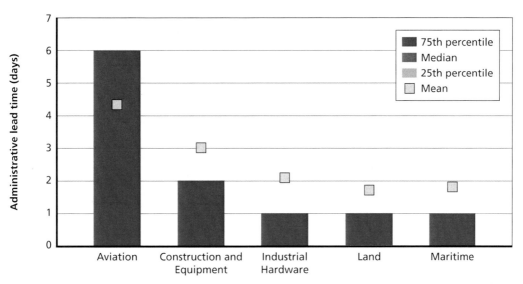

NOTE: 25th percentile and median ALTs are 0 days for automated LTC orders for all five supply chains.

Figure G.5
Administrative Lead Time Statistics for Automated Non-LTC Orders, by Supply Chain

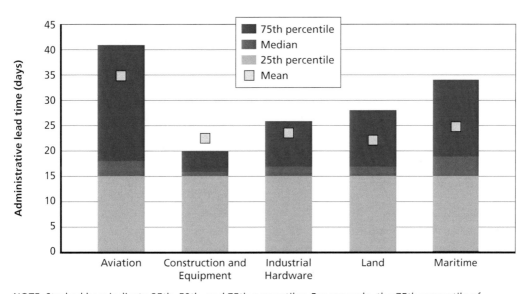

NOTE: Stacked bars indicate 25th, 50th, and 75th percentiles. For example, the 75th percentile of
Construction and Equipment is 20 days; that is, 75 percent of the lead times are less than 20 days.

Figure G.6
Administrative Lead Time Statistics for Manual Orders, by Supply Chain

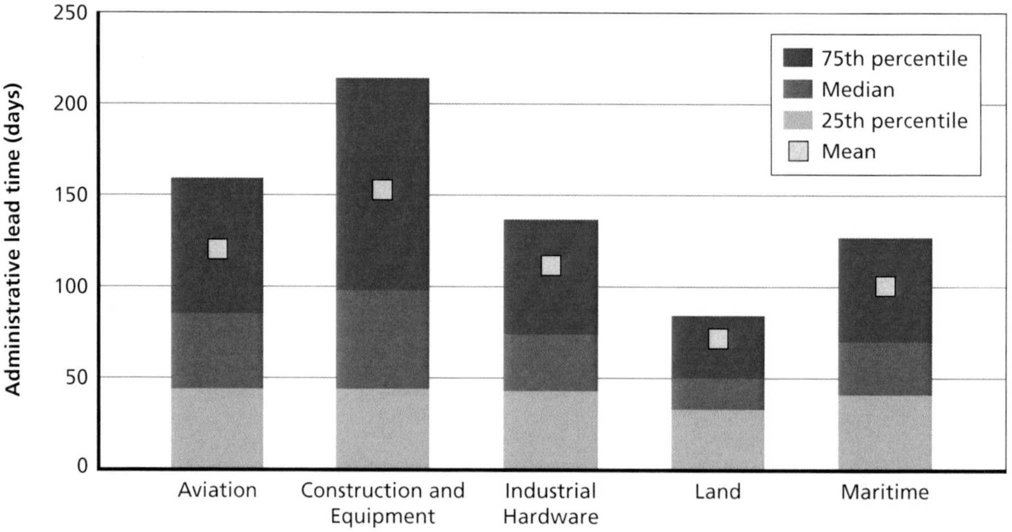

even more pronounced in the top ten percentiles, with both the 90th and 95th percentiles being more than double the typical ALT for other supply chains among automated LTCs and automated non-LTCs. So, the standard process has similar ALTs for LTCs across the supply chains, with Aviation perhaps having more that are nonstandard or have process problems. Note also that the median automated LTC ALT is 0, or same day.

Figure G.6 indicates that Land manual orders showed not only the lowest mean ALT but also the lowest median and 75th-percentile ALTs. On the other end of the spectrum, Construction and Equipment manual orders had higher mean, median, and 75th-percentile times, while Aviation had the second-longest ALTs, both in general and more on the very high end in the right tail of the distribution.

To summarize the ALT statistics by supply chain, automated methods, as expected, have shorter and more-consistent ALTs, with automated LTCs clearly the fastest. Therefore, it would appear that increased use of automation would be preferable, where feasible, assuming that other negative unintended consequences could be avoided. An area to further investigate is why Aviation has more ALTs that do not appear to reflect the speed of the standard automated processes, producing ALTs that are substantially longer. Another area to investigate is why Land's manual ALTs are lower.

Figures G.7, G.8, and G.9 next give PLT percentiles by supply chain for auto-mated LTCs, automated non-LTCs, and manual orders, respectively.[3] A quick review of ALT and PLT figures highlights the substantial percentage of time that PLT accounts for among automated orders. For example, among automated LTCs, PLT accounts for virtually all of the overall wait time associated with an order. Among automated non-LTCs, average supply chain PLT was typically at least twice as long as average supply chain ALT. However, among manual orders, average ALT is actually longer than aver-age PLT for three out of five supply chains.

Aviation PLTs are longer both typically and with more long times in the right tail of the distribution than all other supply chains, for all three major ordering methods. This difference is especially large for automated LTC and manual orders, which together make up more than 80 percent of all Aviation contract lines. Among other supply chains, there were minimal differences in PLTs across the supply chains, with the exception of Construction and Equipment having lower median and average PLTs for automated LTCs. Although the automated LTCs for this supply chain were the most-infrequent lines (about 12,000) among all 15 groups, it would be of interest to further investigate the root cause of this behavior.

Figure G.7
Production Lead Time Statistics for Automated LTC Orders, by Supply Chain

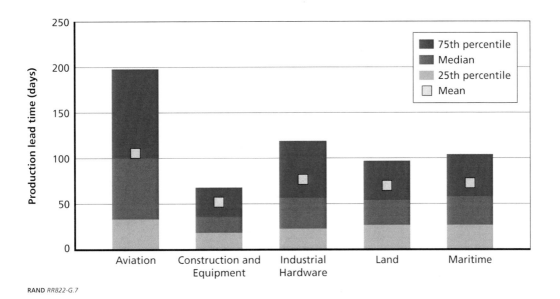

RAND RR822-G.7

---

3   Similar to ALTs, we applied right-censors to PLT data. We censored automated LTC orders at 200 days, auto-mated non-LTC orders at 180 days, and manual orders at 600 days.

Figure G.8
Production Lead Time Statistics for Automated Non-LTC Orders, by Supply Chain

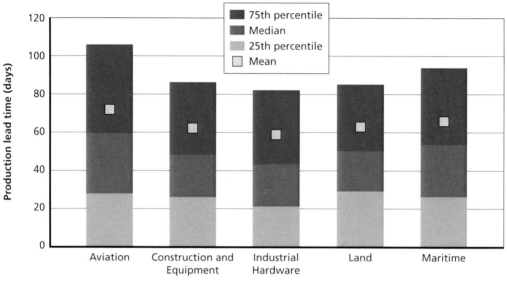

RAND RR822-G.8

Figure G.9
Production Lead Time Statistics for Manual Orders, by Supply Chain

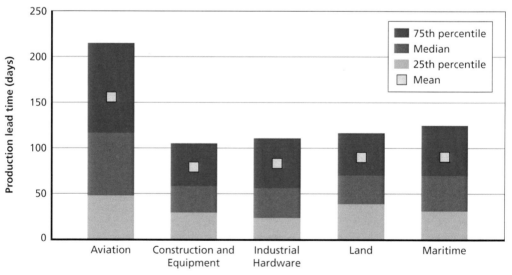

RAND RR822-G.9

Analysis by Federal Supply Group

We wanted to explore how much of the differences in ALT and PLT could be explained by differences in the characteristics of the NIINs. We grouped the NIINs using the FSG, which are the first two digits of the National Stock Number. While this was not a perfect method for identifying common characteristics because differences among NIINs within an FSG could potentially be great, it was a good first step. In the next section, we show how the supply chains compare within FSGs. We focused on FSGs with more than 10,000 contract lines (containing ALT information) between September 2010 and August 2012. As Table G.1 indicates, 16 FSGs made up about 1.11 million contract lines, more than 90 percent of the approximately 1.22 million lines with ALT information that met the filters described initially in this appendix.

Table G.1
Top Federal Supply Groups, by Number of Contract Lines

FSG ID	FSG Label	Number of Contract Lines
53	Hardware and abrasives	326,458
59	Electrical/electronic equipment components	180,618
47	Pipe, tubing, hose, and fittings	115,959
16	Aircraft components/accessories	57,740
15	Aircraft/airframe structure components	55,700
25	Vehicular equipment components	53,004
48	Valves	45,577
61	Electric wire, power distribution equipment	44,297
31	Bearings	39,238
29	Engine accessories	35,076
66	Instruments and laboratory equipment	33,731
30	Mechanical power transmission equipment	32,107
28	Engines, turbines, and components	29,131
43	Pumps and compressors	25,962
62	Lighting fixtures, lamps	22,985
10	Weapons	13,927

NOTE: Table displays the amount of contract lines during September 2010 through August 2012.

Among the 16 FSGs, we counted the number of unique NIINs that were ordered using automated LTCs, automated non-LTCs, and manual approaches. Figure G.10 provides the percentage breakdown of the three ordering methods among FSGs. Clearly, some FSGs rely more on manual orders (e.g., aircraft components/accessories, aircraft/airframe structure components, bearings, and weapons) than others. This is likely related to NIIN characteristics such as demand levels. For NIINs ordered by automated methods, FSGs also varied in the percentages on LTCs.

Figure G.10
Percentage of Unique NIINs Ordered, by Order Method and Federal Supply Group

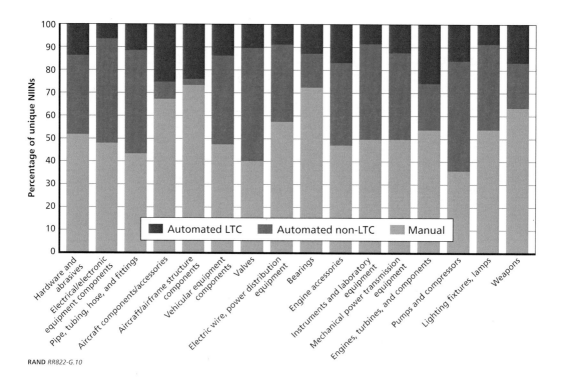

RAND RR822-G.10

Figure G.11 indicates ALT and PLT variation by FSG for automated LTCs. While little variation is present in median ALT, median PLT ranges from around 50 days (for hardware and abrasives, electrical/electronic equipment components, and pipe, tubing, hose, and fittings) to about 100–200 days (aircraft components/accessories, valves, bearings, and engines and turbines). Figure G.12 gives similar information for automated non-LTCs, showing very little variation in median ALT and fairly minimal variation in median PLT (with the exception of aircraft components/accessories and aircraft/airframe structural components). Figure G.13 provides median ALT and PLT for manual orders, demonstrating far more variability among FSGs for both median ALT and PLT than either of the two automated approaches.

Figure G.11
Median Administrative and Production Lead Times for Automated LTC Orders, by Federal Supply Group

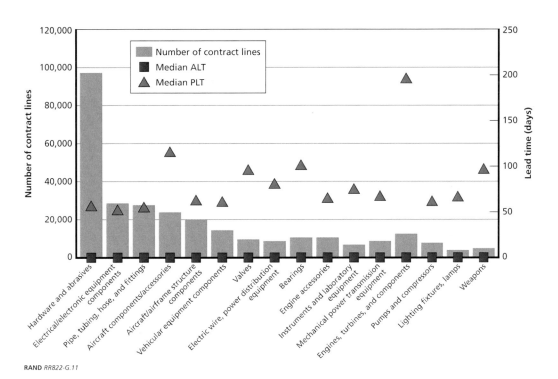

RAND *RR822-G.11*

Figure G.12
Median Administrative and Production Lead Times for Automated Non-LTC Orders, by Federal Supply Group

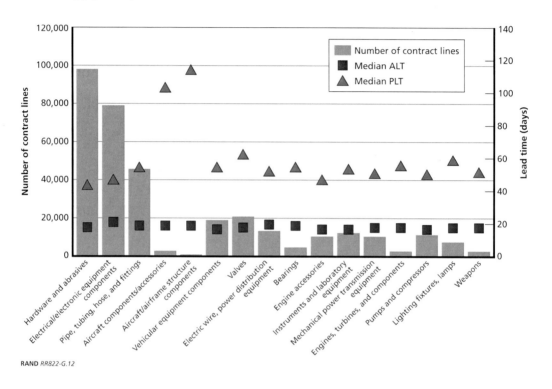

Considering Figures G.11, G.12, and G.13 together, it is clear that median ALT demonstrates far less variability based on FSG than does median PLT, but this is especially the case for automated orders. Table G.2 summarizes the differences in median PLT by FSGs for those that demonstrated the most-extreme tendencies.

Figure G.13
Median Administrative and Production Lead Times for Manual Orders, by Federal Supply Group

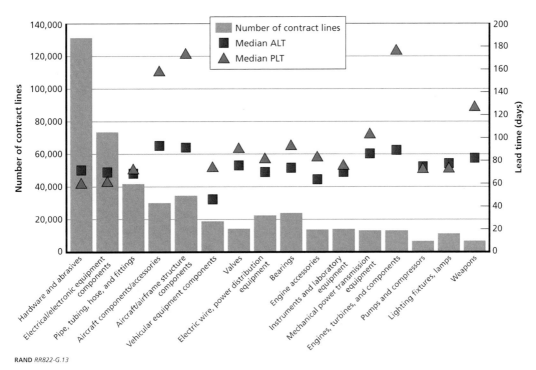

RAND *RR822-G.13*

Table G.2
Differences in Median Production Lead Time, by Federal Supply Group and Order Method

Federal Supply Group	Automated LTC Orders	Automated Non-LTC Orders	Manual Orders
Hardware and abrasives	Lower	Lower	Lower
Electrical/electronic equipment components	Lower		Lower
Pipe, tubing, hose, and fittings	Lower		
Aircraft components/accessories	Higher	Higher	Higher
Aircraft/airframe structure components		Higher	Higher
Valves	Higher		
Bearings	Higher		
Engine accessories		Lower	
Engines, turbines, and components	Higher		Higher
Weapons	Higher		Higher

NOTE: Table displays only those PLTs that demonstrated the most-extreme tendencies.

Analysis by Federal Supply Group and Supply Chain

It is probable that some of the differences in lead times among supply chains come from the effect of NIIN characteristics, hampering the ability to clearly identify the root causes of differences. In this analysis, we attempt to control for NIIN differences by identifying FSGs that were purchased by more than one supply chain. This does not fully control for differences among NIINs because there can still be considerable differences in the characteristics of NIINs within an FSG. We thus analyze performance by supply chain and FSG, again keeping separate automated versus manual and LTC versus non-LTC orders. Because procurement of each NIIN would be assigned to only one supply chain, a direct comparison of the same NIIN purchased by multiple supply chains is not possible. It would take such natural experiments to definitively disentangle the effects.

Table G.3 suggests that even for automated LTCs, although the median ALTs are similar across supply chains, the right tail of the ALT distribution is higher for the Aviation supply chain after accounting for FSG. Table G.4 displays the exact same pattern for automated non-LTCs, with Aviation's 90th and 95th percentiles substantially longer than other supply chains (on a consistent basis). Can this entire trend be explained by aspects of a select number of Aviation NIINs that make long ALTs unavoidable? It is difficult to say without further investigation.

Figures G.14 and G.15 show the ALT comparison for these FSGs in graphical format.

Table G.3
Comparison of Administrative Lead Time for Federal Supply Groups with More Than 500 Automated LTC Lines in Multiple Supply Chains

FSG Label	Supply Chain	Number of Contract Lines	Median ALT	90th Percentile of ALT	95th Percentile of ALT
Engines and turbines and component	Aviation	11,344	1	32	59
Engines and turbines and component	Land	1,515	0	2	8
Engine accessories	Aviation	4,013	1	27	58
Engine accessories	Land	6,440	0	5	11
Mechanical power transmission equipment	Aviation	4,123	0	16	28
Mechanical power transmission equipment	Land	2,849	0	2	6
Mechanical power transmission equipment	Maritime	1,721	0	8	22

Table G.3—Continued

FSG Label	Supply Chain	Number of Contract Lines	Median ALT	90th Percentile of ALT	95th Percentile of ALT
Pumps and compressors	Aviation	654	0	21	62
Pumps and compressors	Maritime	6,661	0	5	14
Pipe, tubing, hose, and fittings	Aviation	5,279	1	21	49
Pipe, tubing, hose, and fittings	Land	2,900	0	3	7
Pipe, tubing, hose, and fittings	Maritime	19,957	0	2	10
Valves	Aviation	2,281	0	15	22
Valves	Land	621	0	2	6
Valves	Maritime	7,161	0	3	13
Hardware and abrasives	Aviation	18,426	0	13	27
Hardware and abrasives	Industrial Hardware	74,507	0	7	19
Hardware and abrasives	Land	2,902	0	3	7
Hardware and abrasives	Maritime	1,106	1	15	43
Electrical/electronic equipment components	Aviation	6,218	1	34	63
Electrical/electronic equipment components	Land	1,238	0	2	4
Electrical/electronic equipment components	Maritime	20,721	0	6	15
Electric wire, power distribution equipment	Aviation	4,361	0	27	48
Electric wire, power distribution equipment	Land	2,717	0	13	30
Electric wire, power distribution equipment	Maritime	1,129	0	19	50
Lighting fixtures, lamps	Aviation	3,312	1	20	36
Lighting fixtures, lamps	Construction and Equipment	703	0	6	16
Instruments and laboratory equipment	Aviation	4,828	0	19	37
Instruments and laboratory equipment	Maritime	1,543	0	7	18

Table G.4
Comparison of Administrative Lead Time for Federal Supply Groups with More Than 500 Automated Non-LTC Lines in Multiple Supply Chains

FSG Label	Supply Chain	Number of Contract Lines	Median ALT	90th Percentile of ALT	95th Percentile of ALT
Engines and turbines and component	Aviation	504	19	95	125
Engines and turbines and component	Land	2,441	17	41	45
Engine accessories	Aviation	1,408	18	82	119
Engine accessories	Land	9,253	17	40	44
Mechanical power transmission equipment	Land	5,150	17	41	47
Mechanical power transmission equipment	Maritime	5,114	18	42	57
Pipe, tubing, hose, and fittings	Land	4,276	17	35	41
Pipe, tubing, hose, and fittings	Maritime	41,252	19	43	54
Valves	Land	705	17	35	41
Valves	Maritime	20,349	18	39	45
Hardware and abrasives	Aviation	7,460	18	82	110.5
Hardware and abrasives	Industrial Hardware	84,787	17	42	59
Hardware and abrasives	Land	4,794	18	36	42
Hardware and abrasives	Maritime	1,118	21	55	90
Electrical/electronic equipment components	Aviation	10,617	20	90	117
Electrical/electronic equipment components	Land	2,046	17	35	41
Electrical/electronic equipment components	Maritime	65,935	21	40	43
Electric wire, power distribution equipment	Aviation	4,332	18	90	122
Electric wire, power distribution equipment	Land	3,165	20	42	48
Electric wire, power distribution equipment	Maritime	6,076	21	40	45
Instruments and laboratory equipment	Aviation	8,082	17	82	123
Instruments and laboratory equipment	Land	637	17	41	47
Instruments and laboratory equipment	Maritime	3,554	18	40	43

Figure G.14
Comparison of Administrative Lead Time for Federal Supply Groups with More Than 500 Automated LTC Lines in Multiple Supply Chains

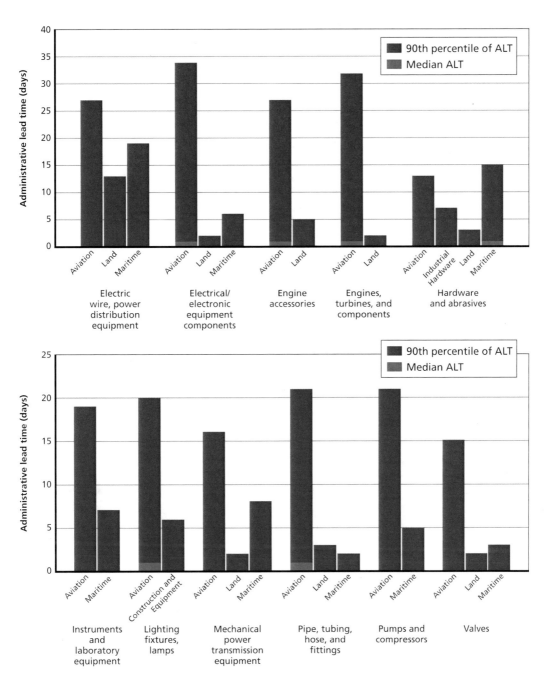

Figure G.15
Comparison of Administrative Lead Time for Federal Supply Groups with More Than 500
Automated Non-LTC Lines in Multiple Supply Chains

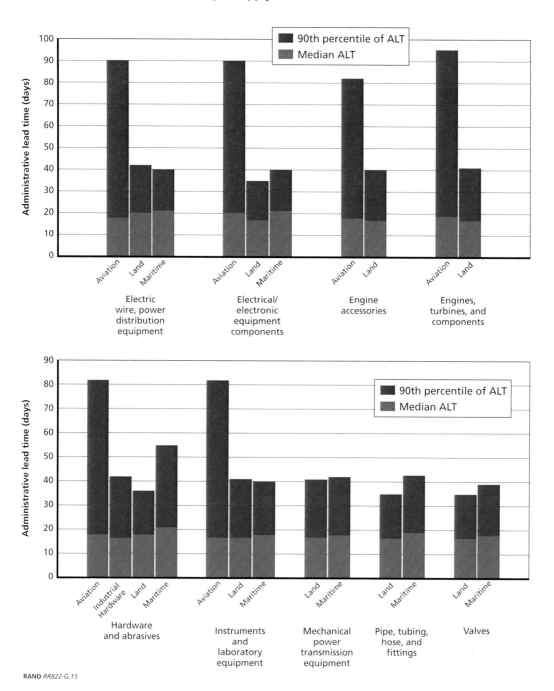

Table G.5 then presents the parallel ALT data for the manual line group, with Aviation's statistics indicating a markedly different pattern of variation from other supply chains compared with Tables G.3 and G.4. Here, while the right tails of their distributions by FSG are still substantially longer than other supply chains, *the medians are also markedly longer*, often more than 20–30 percent longer. The Land supply chain also stands out in the other direction among manual orders, with drastically lower ALTs than other supply chains in many cases. Figure G.16 shows the ALT comparison for these FSGs in graphical format.

Table G.5
Comparison of Administrative Lead Times for Federal Supply Groups with More Than 500 Manual Lines in Multiple Supply Chains

FSG Description	Supply Chain	Number of Contract Lines	Median ALT	90th Percentile of ALT	95th Percentile of ALT
Engines and turbines and component	Aviation	11,523	94	253	327
Engines and turbines and component	Land	1,610	57.5	151.5	194
Engine accessories	Aviation	7,059	91	272	357
Engine accessories	Land	6,738	49	127	166
Mechanical power transmission equipment	Aviation	3,420	97	309.5	436.5
Mechanical power transmission equipment	Land	4,221	73	229	332
Mechanical power transmission equipment	Maritime	5,394	92	245	301
Pumps and compressors	Aviation	1,186	118.5	417	558
Pumps and compressors	Maritime	5,389	69	186	231
Pipe, tubing, hose, and fittings	Aviation	3,957	93	321	425
Pipe, tubing, hose, and fittings	Land	3,918	42	93	119
Pipe, tubing, hose, and fittings	Maritime	33,929	71	196	257
Valves	Aviation	1,565	91	326	431
Valves	Land	507	48	107	130
Valves	Maritime	12,238	77	223	289
Maintenance/repair shop equipment	Aviation	1,560	81	229.5	290
Maintenance/repair shop equipment	Land	596	47	111	130
Hardware and abrasives	Aviation	27,833	79	237	325

Table G.5—Continued

FSG Description	Supply Chain	Number of Contract Lines	Median ALT	90th Percentile of ALT	95th Percentile of ALT
Hardware and abrasives	Industrial Hardware	96,661	74	247	348
Hardware and abrasives	Land	5,161	40	84	102
Hardware and abrasives	Maritime	1,566	76	286	399
Electrical/electronic equipment components	Aviation	19,959	83	255	337
Electrical/electronic equipment components	Land	1,968	44	97	127
Electrical/electronic equipment components	Maritime	51,855	67	208	286
Electric wire, power distribution equipment	Aviation	10,515	85	266	358
Electric wire, power distribution equipment	Land	4,811	67	205	299
Electric wire, power distribution equipment	Maritime	7,180	61	199	265
Lighting fixtures, lamps	Aviation	8,965	81	240	310
Lighting fixtures, lamps	Construction and Equipment	1,437	75	279	343
Instruments and laboratory equipment	Aviation	11,631	71	243	339
Instruments and laboratory equipment	Land	782	68	193	259
Instruments and laboratory equipment	Maritime	1,981	66	187	244

Figure G.16
Comparison of Administrative Lead Times for Federal Supply Groups with More Than 500 Manual Lines in Multiple Supply Chains

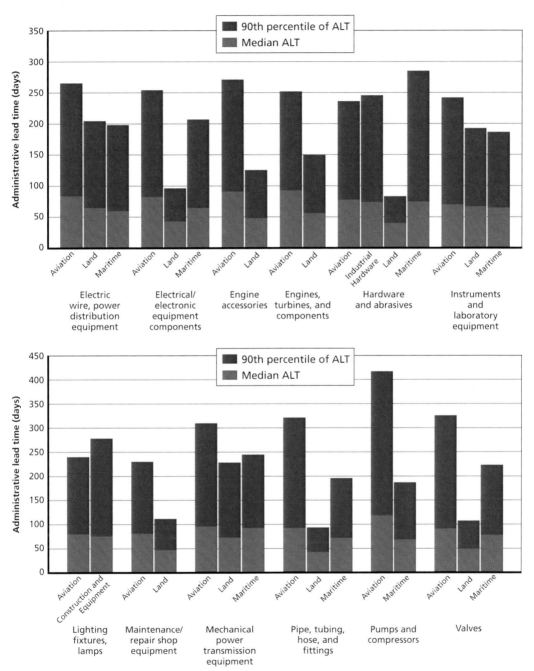

Tables G.6 through G.8 give the same set of data as Tables G.3 through G.5, except now for PLT, with mostly the same supply chains being analyzed by all three major ordering methods.[4] Among automated LTCs, a quick perusal of Table G.6 suggests that Aviation's PLTs are even more consistently out of the ordinary (on a percentage basis relative to other supply chains, as well as from a raw day difference perspective) than were manual PLTs within FSGs. Automated non-LTCs (Table G.7) then present a different picture from any previously described table, with no obvious pattern that suggests that lead times differ by supply chain. Finally, Table G.8 presents manual contract lines, with Aviation once again having significantly longer PLTs and Land demonstrating substantially shorter PLTs (similar to Table G.5). Figures G.17 through G.19 show the PLT comparison for these FSGs in graphical format.

Table G.6
Comparison of Production Lead Times for Federal Supply Groups with More Than 500 Automated LTC Lines in Multiple Supply Chains

FSG Description	Supply Chain	Number of Contract Lines	Median PLT	90th Percentile of PLT	95th Percentile of PLT
Engines and turbines and component	Aviation	7,462	236	495	581
Engines and turbines and component	Land	1,437	36	173	206
Engine accessories	Aviation	3,179	161	323	398
Engine accessories	Land	6,252	42	136	184
Mechanical power transmission equipment	Aviation	2,721	176	383	441
Mechanical power transmission equipment	Land	2,758	36	152	237
Mechanical power transmission equipment	Maritime	1,533	65	245	314
Pumps and compressors	Aviation	520	186.5	297.5	348
Pumps and compressors	Maritime	6,449	60	142	183
Pipe, tubing, hose, and fittings	Aviation	4,274	112	335	400
Pipe, tubing, hose, and fittings	Land	2,807	49	103	119
Pipe, tubing, hose, and fittings	Maritime	19,043	57	142	182
Valves	Aviation	1,756	193	341	423
Valves	Land	603	53	119	134
Valves	Maritime	6,653	85	182	225

---

[4] The criteria to have a minimum of 500 lines with lead time data removed a handful of FSG-supply chain combinations that were present in the parallel ALT charts due to some lines having ALT but not PLT data. Slightly more lines overall had ALT data (1.22 million) than PLT data (1.11 million).

Table G.6—Continued

FSG Description	Supply Chain	Number of Contract Lines	Median PLT	90th Percentile of PLT	95th Percentile of PLT
Hardware and abrasives	Aviation	15,453	92	304	392
Hardware and abrasives	Industrial Hardware	69,028	56	200	266
Hardware and abrasives	Land	2,862	43	90	117
Hardware and abrasives	Maritime	1,088	68	131	147
Electrical/electronic equipment components	Aviation	5,530	69	261	325
Electrical/electronic equipment components	Land	1,205	42	112	145
Electrical/electronic equipment components	Maritime	19,677	53	175	222
Electric wire, power distribution equipment	Aviation	3,581	118	325	412
Electric wire, power distribution equipment	Land	2,562	65	176	211
Electric wire, power distribution equipment	Maritime	1,075	67	203	249
Lighting fixtures, lamps	Aviation	2,995	91	260	293
Lighting fixtures, lamps	Construction and Equipment	662	29	103	147
Instruments and laboratory equipment	Aviation	4,274	105	288	347
Instruments and laboratory equipment	Maritime	1,512	34	104	134

Table G.7
Comparison of Production Lead Times for Federal Supply Groups with More than 500 Automated Non-LTC Lines in Multiple Supply Chains

FSG Description	Supply Chain	Number of Contract Lines	Median PLT	90th Percentile of PLT	95th Percentile of PLT
Engine accessories	Aviation	1,205	110	215	260
Engine accessories	Land	9,093	43	119	148
Mechanical power transmission equipment	Land	4,967	42	133	168
Mechanical power transmission equipment	Maritime	4,831	60	163	210
Pipe, tubing, hose, and fittings	Land	4,180	49	111	134
Pipe, tubing, hose, and fittings	Maritime	40,020	56	145	183

Table G.7—Continued

FSG Description	Supply Chain	Number of Contract Lines	Median PLT	90th Percentile of PLT	95th Percentile of PLT
Valves	Land	682	45	110	137
Valves	Maritime	19,344	64	153	194
Hardware and abrasives	Aviation	7,106	53	144	182
Hardware and abrasives	Industrial Hardware	81,484	43	134	180
Hardware and abrasives	Land	4,675	48	111	131
Hardware and abrasives	Maritime	1,096	49	97	124
Electrical/electronic equipment components	Aviation	10,024	46	133	168
Electrical/electronic equipment components	Land	2,019	48	113	137
Electrical/electronic equipment components	Maritime	64,000	49	141	175
Electric wire, power distribution equipment	Aviation	3,990	56	147	184
Electric wire, power distribution equipment	Land	3,067	55	143	181
Electric wire, power distribution equipment	Maritime	5,847	50	163	206
Instruments and laboratory equipment	Aviation	7,739	57	153	192
Instruments and laboratory equipment	Land	617	48	129	167
Instruments and laboratory equipment	Maritime	3,479	47	126	155

Table G.8
Comparison of Production Lead Times for Federal Supply Groups with More Than 500 Manual Lines in Multiple Supply Chains

FSG Description	Supply Chain	Number of Contract Lines	Median PLT	90th Percentile of PLT	95th Percentile of PLT
Engines and turbines and component	Aviation	8,909	209	534	651
Engines and turbines and component	Land	1,498	70	187	251
Engine accessories	Aviation	5,937	147	361	454
Engine accessories	Land	6,493	57	148	188
Mechanical power transmission equipment	Aviation	2,707	199	505	699
Mechanical power transmission equipment	Land	3,869	69	217	264

Table G.8—Continued

FSG Description	Supply Chain	Number of Contract Lines	Median PLT	90th Percentile of PLT	95th Percentile of PLT
Mechanical power transmission equipment	Maritime	4,504	109	294	373
Pumps and compressors	Aviation	988	134	383	491
Pumps and compressors	Maritime	5,034	68	176	218
Pipe, tubing, hose, and fittings	Aviation	3,132	176	440	531
Pipe, tubing, hose, and fittings	Land	3,723	57	129	166
Pipe, tubing, hose, and fittings	Maritime	30,763	75	194	257
Valves	Aviation	1,334	154	344	438
Valves	Maritime	10,939	91	214	272
Maintenance/repair shop equipment	Aviation	1,332	93.5	244	306
Maintenance/repair shop equipment	Land	562	67	168	197
Hardware and abrasives	Aviation	24,125	89	266	355
Hardware and abrasives	Industrial Hardware	90,233	56	187	247
Hardware and abrasives	Land	4,993	57	129	159
Hardware and abrasives	Maritime	1,505	54	127	161
Electrical/electronic equipment components	Aviation	17,919	74	234	321
Electrical/electronic equipment components	Land	1,907	56	125	148
Electrical/electronic equipment components	Maritime	48,647	61	176	222
Electric wire, power distribution equipment	Aviation	9,057	108	314	406
Electric wire, power distribution equipment	Land	4,481	65	181	222
Electric wire, power distribution equipment	Maritime	6,722	72	201	251
Lighting fixtures, lamps	Aviation	8,248	83	228	308
Lighting fixtures, lamps	Construction and Equipment	1,346	51	147	224
Instruments and laboratory equipment	Aviation	10,372	89	257	316
Instruments and laboratory equipment	Land	690	72.5	184	230
Instruments and laboratory equipment	Maritime	1,863	51	160	197

Figure G.17
Comparison of Production Lead Times for Federal Supply Groups with More Than 500
Automated LTC Lines in Multiple Supply Chains

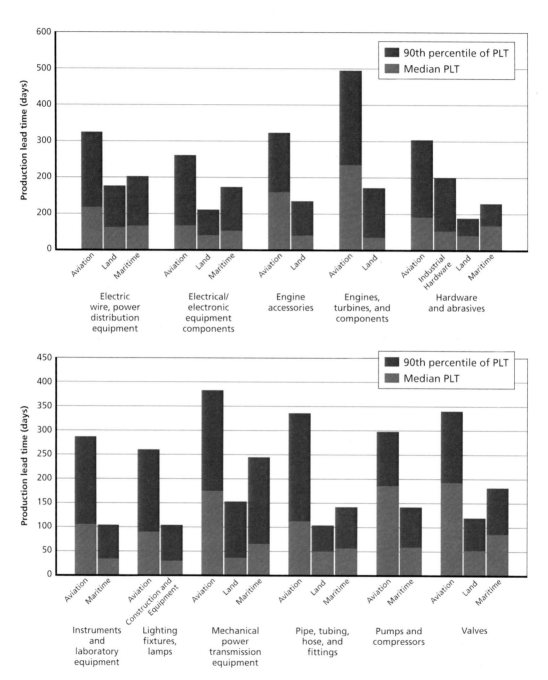

**Figure G.18
Comparison of Production Lead Times for Federal Supply Groups with More Than 500
Automated Non-LTC Lines in Multiple Supply Chains**

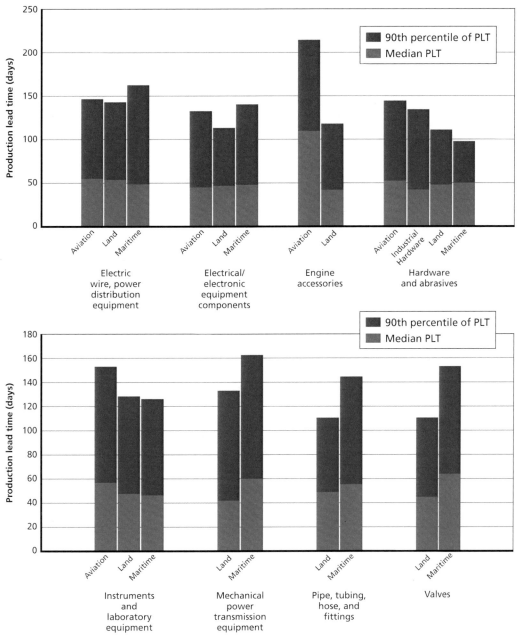

Figure G.19
Comparison of Production Lead Times for Federal Supply Groups with More Than 500
Manual Lines in Multiple Supply Chains

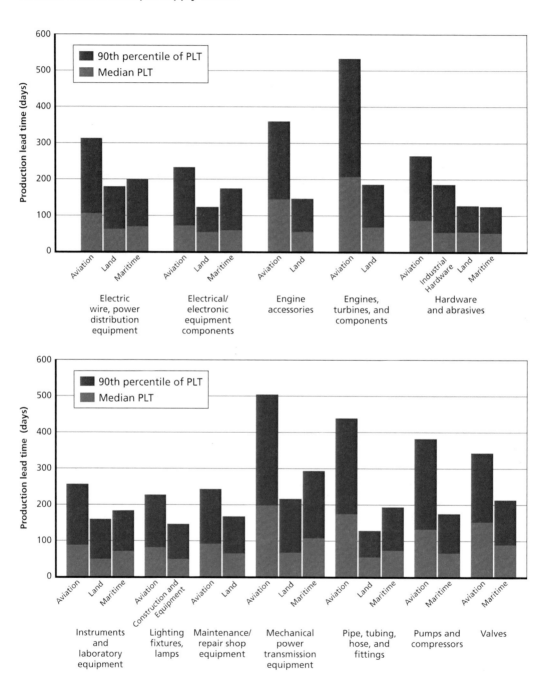

Regression Analysis Examining More Variables

We wanted to see whether other factors associated with the NIIN or the size of the order would explain some of the differences in ALT and PLT. We conducted a multiple regression analysis to determine the effect on ALT and PLT of the factors discussed above—namely, automation, LTC, supply chain, and FSG—as well as these additional factors:

- NIIN awarded or not in last 12 months, reflecting recent activity
- order quantity
- unit cost
- unit weight
- contract dollar range.

Some variables in the ALT data set and the PLT data set were right-censored at their 95th percentile (Table G.9).[5] We decided to do this because response times and the continuously distributed predictor variables for lead time data are extremely right-skewed (i.e., not normal), which presents some complications when interpreting statistical tests designed for responses with normal distributions. While transforming the response variable can help address this issue, data points in the far right tail of distributions (either for independent or dependent variables) can still have a disproportionate influence on the fit of a regression analysis if left unmodified. Given the nature of this project—studying ALT and PLT in general as a function of various predictors—right-censoring values represented a simple and reasonable compromise between leaving data unaltered and throwing out outliers completely.

Table G.9
Percentiles of Interest, by Variable

Variable Summarized	10th Percentile	Median	90th Percentile	95th Percentile
ALT	0	26	150	224
PLT	14	64	212	290
Unit cost	$2.20	$69.80	$1,364.90	$2,852.70
Unit weight	0.01	0.3	9.0	25.0
Order quantity	2	19	369	956

NOTE: Some variables in the ALT data set and PLT data set were right-censored at their 95th percentile.

---

5   ALT, unit cost, unit weight, and order quantity were summarized using the data set with ALT values, whereas PLT was summarized using only the data set with PLT values. Although technically the 95th percentile was slightly different for unit cost, unit weight, and order quantity when using the ALT data set versus the PLT data set, the differences were minimal. For these three variables, we used the ALT data set for the official right-censor because it contained slightly more data points (1.22 million compared to 1.11 million).

Before we attempt to explain regression analysis with many predictors, we provide a simple example to clarify the nature of the regression methods that we used. The beginning stages of regression analysis on ALT simply estimated the effect of the three major order methods on the estimate of average ALT. A multiple linear regression model was fit relating ALT to the three order methods, where we used the natural logarithm of (ALT + 1) as the response variable and treated the automated non-LTC lines as the *baseline* group (i.e., where manual_group = 0 and LTC_group = 0).[6] The resulting equation is

$$\ln(\text{ALT} + 1) = 2.98325 + 1.37266(\text{manual\_group}) - 2.14187(\text{LTC\_group}). \quad \text{(G.1)}$$

Here, the interpretation is that $e^{\wedge}(1.37266) = 3.9453 =$ the ratio of the expected geometric mean of (ALT + 1) for the manual lines over the expected geometric mean of (ALT + 1) for the automated non-LTC lines.[7] In other words, ALT times were fit in this model as being nearly 400 percent higher for manual lines compared with automated non-LTC lines.

Similarly, $e^{\wedge}(-2.14187) = 0.1175$ indicates that automated LTCs have ALTs approximately 90 percent lower than automated non-LTCs. (Recall Figure G.1 in which the medians somewhat reflect these parameters.)

Initial regression analysis on PLT was performed using a similar approach, with the following fitted equation:

$$\ln(\text{PLT} + 1) = 3.86554 + 0.39982(\text{manual\_group}) + 0.23258(\text{LTC\_group}). \quad \text{(G.2)}$$

Automated LTCs had PLTs that were about 26 percent higher than automated non-LTCs, and manual lines' PLTs were about 49 percent higher than automated non-LTCs, which matches up fairly closely with Figure G.2. While these numbers are significant from a predictive perspective (as judged by individual predictor p-values of less than .0001), we should note that the adjusted R-squared value for the model fit shown in Equation G.1 was much higher than for the model fit in Equation G.2 (0.71 versus 0.03). In summary, whether using informal or formal methods, the effect of automated versus manual and LTC versus non-LTC was extremely significant from a predictive perspective for ALT but much less so for PLT.

We performed similar analysis using multiple predictors in models, ultimately determining a rank of importance relative to other variables on ALT and PLT. Inter-

---

[6]   We used the natural logarithm of (ALT+1) for some models to help alleviate the extreme nonnormality of ALT. Note that the integer 1 had to be added to ALT times because they are equal to zero in many cases and the natural log function is invalid at zero. For more on how to interpret log-response model parameters, see Andrew Gelman and Jennifer Hill, *Data Analysis Using Regression and Multilevel/Hierarchical Models*, New York: Cambridge University Press, 2007.

[7]   The *geometric mean* is the *n*th root of the product of *n* numbers. It is thus different from the more commonly seen *arithmetic mean*, which is the sum of *n* numbers divided by *n*.

preting the relative importance of variables becomes more challenging with many predictors, in part due to multicollinearity among several predictors. Additionally, some variables appear closely correlated with ALT or PLT for only certain subsets of data (such as the effect of FSG on ALT for manual orders but not automated orders). For these reasons, we caution against precise interpretation of regression parameters and instead offer a general ranking of variables into groups affecting ALT and PLT. Figure G.20 depicts the rankings for ALT, plus conceptual interpretations of the effects. Here, the lines shaded in red affect ALT the most, as prior figures in this appendix would suggest. Purple rows represent significant variables that have a moderate effect—not as major as manual versus automated and LTC versus non-LTC, but still substantial. Blue rows indicate variables that statistically show as being significant, but on a minor scale.

Figure G.21 offers a parallel view of variable importance for PLT. Among the most notable differences are the elevated effects of supply chain and FSG, as well as the diminished importance of a contract line being part of an LTC.

Figure G.20
Variable Importance on Administrative Lead Time

Variable	Illustration of Effect on PLT
Manual versus automated	Automated is 78 days faster, on average.
LTC versus non-LTC	Automated LTCs were 20 days faster than automated non-LTCs, on average.
NIIN awarded in last 12 months (versus not)	After accounting for manual versus automated and LTC versus non-LTC, average of 18 days faster if NIIN was recently awarded.
Supply chain	Fast to slow (when adjusting for manual, automated LTC, and automated non-LTC): Land, Maritime, Industrial Hardware, Construction and Equipment, Aviation. The difference between Land and Aviation was 20 days, on average.
Federal Supply Group	Clear relationship, especially on manual orders.
Order quantity	Slight positive correlation with ALT.
Unit cost	Minimal positive correlation with ALT, except in the right tail, where it becomes more significant.
Contract dollar range	ALTs were about 10 days faster when contract value was less than $150,000 versus more than $150,000.

NOTE: Factors are listed in decreasing order of importance. Red = major; purple = moderate; blue = minor.
RAND RR822-G.20

Figure G.21
Variable Importance on Production Lead Time

Variable	Illustration of Effect on PLT
Manual versus automated	Automated is 25 days faster, on average.
Supply chain	Fast to slow (when adjusting for manual, automated LTC, and automated non-LTC): Construction and Equipment, Industrial Hardware, Land, Maritime, and Aviation. Aviation was about 47 days higher than Construction and Equipment. Industrial Hardware, Land, and Maritime were all within ten days of Construction and Equipment.
Federal Supply Group	FSG explains more of the variability in PLT than in ALT, despite PLT being more variable in general.
Contract dollar range	PLTs were about 30 days faster when contract value was less than $150,000.
Unit cost	There is a more moderate positive correlation with ALT, but with PLT, the effect is still more significant in the far right tail.
LTC versus non-LTC	Medians indicate that LTC is 13 days faster.
NIIN awarded in last 12 months (versus not)	PLT was about six days faster on average if a NIIN was recently awarded.
Order quantity	When accounting for all other significant variables, the difference between exactly one item and between 100 and 500 items was approximately 35 days, but the effect appeared small in models with fewer predictors.

NOTE: Factors are listed in decreasing order of importance. Red = major; purple = moderate; blue = minor.
RAND RR822-G.21

Use of Automation and LTC by Supply Chain

This section provides supplementary visuals on the differences among supply chains and order methods, which may be fully or mostly driven by NIIN characteristics instead of differences in practices among the supply chains. Figure G.22 indicates the percentage of contract lines by supply chain and order group, with Aviation having the highest percentage of manual lines and of automated LTCs. Construction and Equipment and Maritime had the highest percentage of automated non-LTCs. Land, Maritime, and Industrial Hardware all eclipsed 60 percent for their total automated lines percentage.

Figure G.23 depicts the percentage of unique NIINs ordered by supply chain for each order method. This figure appears proportionally similar to Figure G.22 in terms of the percentage split of order method by supply chain, although with a slightly higher percentage of manual orders in all supply chains. These two figures suggest that supply chains do not have extreme differences in their mixes of order types, whether examining contract lines or unique NIINs ordered. That being said, for all five supply chains, manual orders do have a higher share of unique NIIN orders relative to share of contract lines. This is expected, given that frequently ordered items are more suited for automated order processes.

Figure G.22
Percentage of Contract Lines, by Supply Chain and Order Method

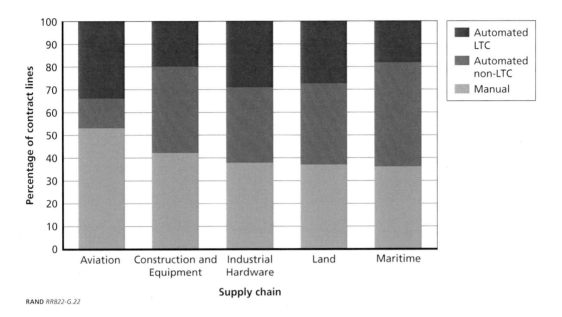

RAND *RR822-G.22*

Figure G.23
Percentage of Unique NIINs Ordered, by Supply Chain and Order Method

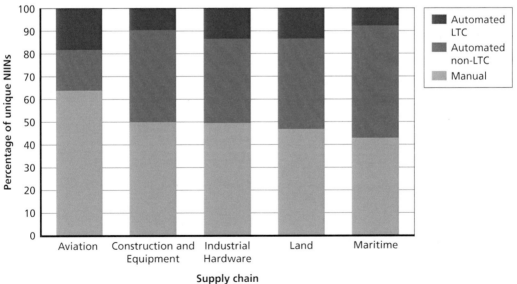

RAND *RR822-G.23*

Figure G.24 graphically presents the percentage of contract value by the three major order methods for all five supply chains, with Land's manual orders composing a larger share proportionally of contract value than did their unique NIINs or contract lines percentages. The automated LTC lines for Construction and Equipment accounted for more than 40 percent of contract value, despite being only about 20 percent of total lines.

Finally, Figures G.25 through G.27 offer more insight into the nature of NIINs purchased by supply chain and order method, using annual demand averages, unit price, and average ADV.

As Figure G.25 shows, for most supply chains, the items being ordered via automated LTC have, on average, higher average annual demand. Construction and Equipment is an exception, where manually ordered items have higher average annual demand.

Figure G.26 presents a comparison of unit price between supply chains and order methods. The chart shows that differences in unit prices lie more with the supply chain, with Aviation's NIINs having much higher unit prices on average and Industrial Hardware falling easily on the low end of the scale. Among ordering methods, there are inconsistent differences in the unit prices of NIINs ordered by automated LTC compared with manual; NIINs ordered by automated non-LTC methods do have lower prices, perhaps in keeping with the cost threshold restrictions for that ordering mechanism.

Figure G.24
Percentage of Contract Value, by Supply Chain and Order Method

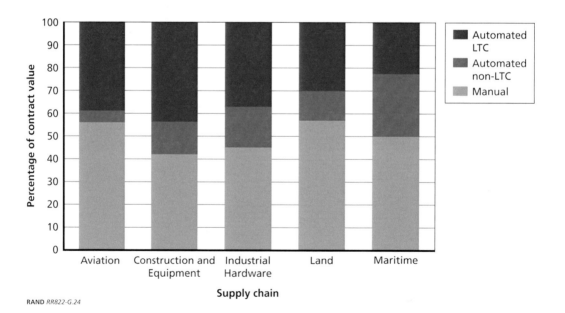

RAND RR822-G.24

**Figure G.25
Average Annual Demand of NIINs Ordered, by Supply Chain and Order Method**

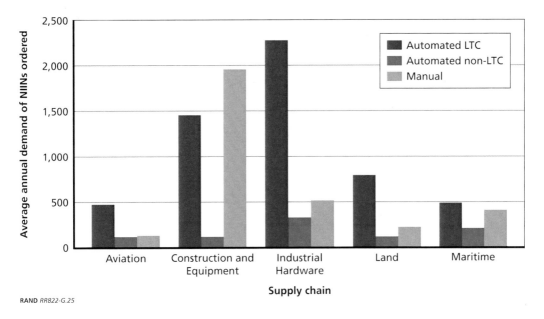

**Figure G.26
Average Unit Price of NIINs Ordered, by Supply Chain and Order Method**

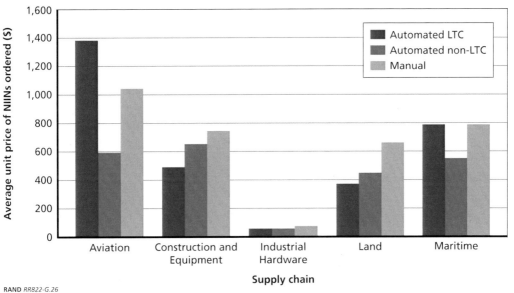

Figure G.27 presents average ADV by supply chain and order type. In each supply chain, the items ordered using automated LTC have, on average, the highest ADVs. This is consistent with what we heard in interviews at supply chains, where it was reported that efforts to place items onto LTCs tend to focus on the high-ADV items. It is also compatible with our own finding that placing high-ADV items onto LTCs would bring the biggest benefit in reducing PR workload and inventory investment.

Figure G.27
Average Annual Demand Value of NIINs Ordered, by Supply Chain and Order Method

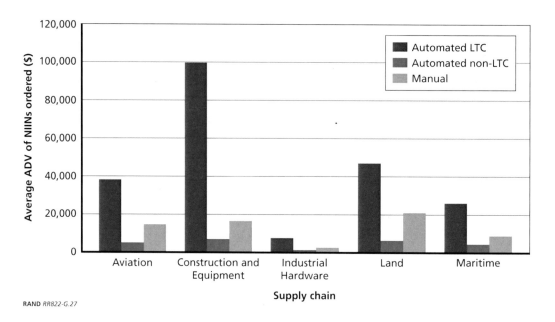

Selecting NIIN Candidates for Long-Term Contracts

One of the strategies for increasing the supply chain agility of DLA is to place regularly purchased items onto LTCs, known in DLA as *outline agreements*. As shown in Figures G.23 and G.24 in Appendix G, purchases of items on LTC accounted for 7–18 percent of NIINs and 22–43 percent of the dollars spent from September 2010 to August 2012, depending on supply chain. This suggests that there are still opportunities to further increase the use of LTCs.

The benefit of an LTC is that DLA does not need to solicit bids and award a contract each time it needs to purchase the item, resulting in a vastly shorter ALT. This shorter lead time enables DLA to be more responsive to changes in customer demand. It is generally easier to forecast demand over shorter time horizons. If the item does experience an increase in demand, shorter lead times enable a faster replenishment of inventory to prevent backorders, or else fill them.

In addition to benefiting the item on contract by reducing its ALT, placing items on LTCs benefits the system as a whole. It reduces the burden on processing PRs by requiring far less labor because delivery orders are often processed electronically through automated systems. This frees up workload among the acquisition personnel to process other PRs.

The second benefit is related to the first. The lower burden on placing delivery orders means that DLA can place more-frequent orders, in smaller quantities. This enables DLA to commit to and store less inventory. As a result, there is less chance of overstocking the item should the demand for the item dramatically decrease, which would eventually lead to disposing the excess inventory at a loss. The capital that is freed by avoiding the purchase of large amounts of LTC items can be redeployed to buy more safety stock of critical items or to reduce total costs.

Soliciting, awarding, and setting up an LTC requires a great deal of time, effort, and expertise. Consequently, DLA should create a prioritization scheme to identify which NIINs to target. In this appendix, we show that placing NIINs with high ADV onto LTCs will best contribute to reducing PR workload and reducing inventory investment. It may be that not all of these NIINs can be placed onto LTCs and that DLA may have other criteria that it must consider to meet operational requirements.

However, selecting NIINs based on ADV will serve as a starting point for prioritizing LTC efforts.

Reducing the Purchase Request Burden

Methods based on EOQ, including the DLA CovDur Table, will lead to ordering high-ADV items in shorter coverage durations (i.e., more frequently) in order to save on inventory costs. Consequently, replenishments for high-ADV items will generate a large number of PRs compared with lower-ADV items. This can be shown mathematically. The number of PRs per year of an item is given by dividing the annual demand rate λ by the order quantity Q. If EOQ is used, the number of PRs per year will be:

$$\text{PR per year} = \frac{\lambda}{Q^*} = \frac{\lambda}{\sqrt{\dfrac{2k\lambda}{hc}}} = \sqrt{\frac{h}{2k}}\sqrt{\lambda c}.$$

Notice that λc is the ADV. The formula shows that when EOQ or the CovDur Table is used, the number of PRs per year for an item is an increasing function of the item's ADV. Items with high ADV will have a large number of PRs per year. The burden of manually processing these PRs can be greatly reduced if these items are moved to LTC with automated ordering.

Reducing Inventory Investment

The average inventory value for an item is given by multiplying the unit cost c by the average inventory on hand, which is half the order quantity, or $Q/2$. Combined with the EOQ formula, the average inventory value is

$$\text{average inventory value} = c\frac{Q^*}{2} = \frac{c}{2}\sqrt{\frac{2k\lambda}{hc}} = \frac{1}{2}\sqrt{\frac{2k}{h}}\sqrt{\lambda c}.$$

The average inventory value for an item when EOQ is used is thus an increasing function of the ADV λc. Thus, items with high ADV are the items for which the inventory investment will be high.

Once these high-ADV items are moved to LTCs, the lower cost of placing delivery orders enables DLA to order them even more frequently, in even smaller batches, enabling still further reduction in inventory. If the EOQ under manual ordering, with a per-PR cost of k_M, is

$$Q_M^* = \sqrt{\frac{2k_M\lambda}{hc}},$$

and the EOQ under automated ordering, with a per-PR cost of k_A, is

$$Q_A^* = \sqrt{\frac{2k_A\lambda}{hc}},$$

then the difference in inventory value is

$$c\frac{Q_A^*}{2} - c\frac{Q_M^*}{2} = \frac{c}{2}\left(\sqrt{\frac{2k_A\lambda}{hc}} - \sqrt{\frac{2k_M\lambda}{hc}}\right) = \frac{c}{2}\sqrt{\frac{2(k_A - k_M)\lambda}{hc}} = \left(\frac{1}{2}\sqrt{\frac{2(k_A - k_M)}{h}}\right)(\sqrt{\lambda c}),$$

which is an increasing function of the ADV λc. Thus, the items with the greatest reduction in inventory investment will be the items with the highest ADVs.

References

Avery, S., "Purchasing Salutes Suppliers; World-Class Companies Use Recognition Programs as a Tool to Develop and Reward Preferred Suppliers," *Purchasing Magazine*, 2008, p. 63.

Brown, Robert G., *Decision Rules for Inventory Management*, New York: Holt, Rinehart and Winston, 1967.

Chenoweth, Mary, Jeremy Arkes, and Nancy Moore, *Best Practices in Developing Proactive Supply Strategies for Air Force Low-Demand Service Parts*, Santa Monica, Calif.: RAND Corporation, MG-858-AF, 2010. As of October 9, 2014:
http://www.rand.org/pubs/monographs/MG858.html

Collins, M., "QRM for Reducing Lead Times," *Industrial Maintenance and Plant Operation*, December 2008.

Daniels, A. K., "Self-Deception and Self-Discovery in Fieldwork," *Qualitative Sociology*, Vol. 6, No. 3, 1983, pp. 195–214.

Day, J., "Suppliers Take Center Stage," *Purchasing Magazine*, 2008, p. 49.

Deemer, R. L., A. J. Kaplan, and W. K. Kruse, *Application of Negative Binomial Probability to Inventory Control*, AMC Inventory Research Office, Institute of Logistics Research ALMC, Philadelphia, Pa., 1974.

Defense Logistics Agency, "First Destination Transportation and Packaging Initiative (FDTPI)," web page, undated. As of October 8, 2014:
http://www.dla.mil/FDTPI/Pages/default.aspx

———, *Procedures, Guidance, and Information (PGI)*, Defense Logistics Acquisition Directive, December 4, 2012.

———, *SMSG Business Rules*, Revision 5, Defense Logistics Acquisition Directive, September 9, 2013.

DLA—*See* Defense Logistics Agency.

DoD—*See* U.S. Department of Defense.

FAR—*See* Federal Acquisition Regulation.

Federal Acquisition Regulation, 48 C.F.R., 2014. As of October 14, 2014:
http://www.acquisition.gov/far/

Flextronics, "Flextronics Receives Emulex Supplier of the Year Award," *PR Newswire*, March 2012.

Gelman, Andrew, and Jennifer Hill, *Data Analysis Using Regression and Multilevel/Hierarchical Models*, New York: Cambridge University Press, 2007.

Gligor, D. M., and M. C. Holcomb, "Antecedents and Consequences of Supply Chain Agility: Establishing the Link to Firm Performance," *Journal of Business Logistics*, Vol. 33, No. 4, December 2012a, pp. 295–308.

———, "Understanding the Role of Logistics Capabilities in Achieving Supply Chain Agility: A Systematic Literature Review," *Supply Chain Management: An International Journal*, Vol. 17, No. 4, 2012b, pp. 438–453.

Gligor, D. M., M. C. Holcomb, and T. P. Stank, "A Multidisciplinary Approach to Supply Chain Agility: Conceptualization and Scale Development," *Journal of Business Logistics*, Vol. 34, June 2013, pp. 94–108.

Goodrich Corporation, "Goodrich Presents US Tool Group with Strategic Supplier Award," *Business Wire*, 2009.

Hadley, G., and T. M. Whitin, *Analysis of Inventory Systems*, Englewood Cliffs, N.J.: Prentice Hall, 1963.

Hafey, S., "Managing Supplier Relationships Is Key to Business Success," *Supply Chain Solutions*, September/October 2010, pp. 36–39.

Harris, F. W., "How Many Parts to Make at Once," *Factory: The Magazine of Management*, Vol. 10, 1913, pp. 135–136, 152.

Headquarters DLA, "Agency Performance Review," briefing, September 2013.

Hiiragi, S., *The Significance of Shortening Lead Time from a Business Perspective*, Discussion Paper Series, No. 391, Manufacturing Management Research Center, University of Tokyo, March 2012.

Hopp, W., M. Spearman, and D. Woodruff, "Practical Strategies for Lead Time Reduction," *Manufacturing Review*, Vol. 3, No. 2, 1990.

Jankowski, N. W., and K. B. Jensen, *A Handbook of Qualitative Methodologies for Mass Communication Research*, New York: Routledge, 1991.

KEMET Corporation, "KEMET Receives Rockwell Collins 2012 Top Supplier Award," *PR Newswire*, 2012.

Krause, Daniel R., and Robert B. Handfield, *Developing a World-Class Supply Base*, Tempe, Ariz.: Center for Advanced Purchasing Studies, 1999.

Lahti, M., A. Shamsuzzoha, and P. Helo, "Developing a Maturity Model for Supply Chain Management," *International Journal of Logistics Systems and Management*, Vol. 5, No. 6, 2009.

Li, X., C. Chung, T. J. Goldsby, and C. W. Holsapple, "A Unified Model of Supply Chain Agility: The Work-Design Perspective," *International Journal of Logistics Management*, Vol. 19, No. 3, 2008, pp. 408–435.

Lockheed Martin, "Lockheed Martin Reaches Strategic Supply Chain Agreement," *ENP Newswire*, 2012.

Mabin, Victoria J., "A Practical Near-Optimal Order Quantity Method," *Engineering Costs and Production Economics*, Vol. 15, 1988, pp. 381–386.

"Making the Connections," *Metalworking Production*, Vol. 153, No. 2, March 2009, p. 44.

Miel, R., "GM 'Metric' System Rates Suppliers," *Plastics News*, Vol. 16, No. 16, 2004, pp. 7–10.

Morris, W., "How to Leverage Supplier Performance Management for Continuous Supply Chain Improvement," *Supply & Demand Chain Executive*, May/June 2010.

Muckstadt, J. A., and A. Sapra, *Principles of Inventory Management: When You Are Down to Four, Order More*, New York: Springer Science + Business Media, 2010.

Office of Management and Budget, *Guidelines and Discount Rates for Benefit-Cost Analysis of Federal Programs*, Circular A-94, October 29, 1992 (Appendix C, revised December 26, 2013). As of March 12, 2014:
http://www.whitehouse.gov/omb/circulars_a094/a94_appx-c

Olson, J., "Lead Times: It's All About the Process," *Surface Fabrication*, September 2009.

"Palletized Production Reduces Lead Times," *Quality*, March 2011, pp. 52–54.

Peltz, E., and M. Robbins, with G. McGovern, *Integrating the Department of Defense Supply Chain*, Santa Monica, Calif.: RAND Corporation, TR-1274-OSD, 2012. As of September 24, 2014:
http://www.rand.org/pubs/technical_reports/TR1274.html

Perry, J. H., "Lead Time Management: Private and Public Sector Practices," *Journal of Purchasing and Materials Management*, September 1990.

Porter, A., "Lead Times Are Shrinking, but Not Everyone's a Winner," *Purchasing Magazine*, 1998.

Presutti, Victor J., Jr., and Richard C. Trepp, *More Ado About Economic Order Quantity (EOQ)*, Operations Research Office, Headquarters, Air Force Logistics Command, 1970.

PRTM Management Consultants, "Supply Chain Management Maturity Model: Understand the Transformation Required to Move from a Functionally Focused Supply Chain to Cross-Enterprise Collaboration," 2005.

Purchasing Magazine Staff, "Spotlight Shines on Key Suppliers," *Purchasing Magazine*, 2009, p. 58.

Reaume, J., "6 Procurement Actions That Can Boost Your Business," *Supply Chain Management Review*, 2010, pp. 48–53.

Schier, T., "Engine Maker P&W Looks to Lock in Material Supplies for 30 Years," *Metal Bulletin*, 2012.

Seidman, I., *Interviewing as Qualitative Research: A Guide for Researchers in Education*, New York: Teachers College Press, 2012.

Spiegel, R., "SRM Leaders Outpace Peers on Lead Times, Other Key Metrics," *Supply Chain Management Review*, January/February 2011, pp. 47–49.

Stank, T., J. P. Dittman, C. Autry, K. Petersen, M. Burnette, and D. Pellathy, "Bending the Chain: The Surprising Challenge of Integrating Purchasing and Logistics," Knoxville, Tenn.: University of Tennessee Global Supply Chain Institute, Spring 2014.

Supply Chain Council, *Supply Chain Operations Reference (SCOR) Model*, Version 10, 2010.

Teague, P., "P&G Is King of Collaboration—for Building Close and Productive Relationships Internally and with Suppliers," *Purchasing Magazine*, 2008, p. 46.

Trent, R., "Creating the Ideal Supplier Scorecard," *Supply Chain Management Review*, March/April 2012.

Trent, R., and R. Monczka, "Purchasing and Supply Management: Trends and Changes Throughout the 1990s," *International Journal of Purchasing and Materials Management*, Fall 1998.

U.S. Department of Defense, *The Defense Acquisition System*, Directive 5000.01, May 12, 2003a.

———, *DoD Supply Chain Materiel Management Regulation*, Office of the Deputy Under Secretary of Defense for Logistics and Materiel Readiness, DoD 4140.1-R, May 23, 2003b.

———, *Comprehensive Inventory Management Improvement Plan*, Office of the Assistant Secretary of Defense for Logistics and Materiel Readiness, October 2010.

———, *DoD Supply Chain Materiel Management Procedures: Demand and Supply Planning*, DoD Manual 4140.01, Vol. 2, February 2014a.

———, *DoD Supply Chain Materiel Management Procedures: Materiel Returns, Retention, and Disposition*, DoD Manual 4140.01, Vol. 6, February 2014b.

U.S. Government Accountability Office, *Defense Inventory: Opportunities Exist to Improve the Management of DOD's Acquisition Lead Times for Spare Parts*, GAO-07-281, Washington, D.C., March 2, 2007a.

———, *Defense Inventory: Opportunities Exist to Save Billions by Reducing Air Force's Unneeded Spare Parts Inventory*, GAO-07-232, Washington, D.C., April 27, 2007b.

———, *Defense Inventory: Management Actions Needed to Improve the Cost Efficiency of the Navy's Spare Parts Inventory*, GAO-09-103, Washington, D.C., December 12, 2008.

———, *Defense Inventory: Army Needs to Evaluate Impact of Recent Actions to Improve Demand Forecasts for Spare Parts*, GAO-09-199, Washington, D.C., January 12, 2009.

———, *Defense Inventory: Defense Logistics Agency Needs to Expand on Efforts to More Effectively Manage Spare Parts*, GAO-10-469, Washington, D.C., May 11, 2010.

———, *DOD's 2010 Comprehensive Inventory Management Improvement Plan Addressed Statutory Requirements, but Faces Implementation Challenges*, GAO-11-240R, Washington, D.C., January 7, 2011.

———, *Defense Inventory: Actions Underway to Implement Improvement Plan, but Steps Needed to Enhance Efforts*, GAO-12-493, Washington, D.C., May 3, 2012.

Weiss, R. S., *Learning from Strangers: The Art and Method of Qualitative Interview Studies*, New York: Simon and Schuster, 1995.

Wilson, R. H., "A Scientific Routine for Stock Control," *Harvard Business Review*, No. 13, 1934, pp. 116–128.

Wouters, M., "Economic Evaluation of Lead Time Reduction," *International Journal of Production Economics*, 1991, pp. 111–120.

Youngman, K. J., "A Guide to Implementing the Theory of Constraints," web page, 2009. As of October 13, 2014:
http://www.dbrmfg.co.nz/Supply%20Chain%20Lead%20Times.htm